What People Are Saying About

Resetting Our Future: Navigating Clean Energy Choices

As a climate leader at the World Wildlife Fund, I know small businesses are essential to a clean energy future – but often lack the tools to navigate the transition. This book delivers exactly what's needed: a practical, clear-eyed roadmap for owners ready to cut costs, reduce risk, and make meaningful progress on sustainability. Through my work with the Montgomery County Green Bank, I've seen how eager small business leaders are to act – yet they're overwhelmed by technical complexity, upfront costs, and fear of missteps. That's why this book is so valuable. It meets business leaders where they are, with a simple, strategic framework that builds confidence and drives action. With the CLEAR approach, it empowers small businesses to become more resilient, competitive, and climate-ready – one smart decision at a time.
Marcene Mitchell, Senior Vice President, Climate Change, WWF; Chair, Montgomery County Green Bank

This book is a goldmine. Elvin sketches the context for business decision-making in an elegant and approachable way and outlines a deceptively simple methodology for weighing options and taking positive action.
Tom Bowman, author of *The Green Edge* and *What If Solving the Climate Crisis Is Simple?*

This book is exceptionally rich in content, well-structured, and serves as an invaluable resource for guiding small businesses through their energy transition journey. I especially appreciated the CLEAR approach, which presents a holistic yet straightforward roadmap. By simplifying complex concepts and illustrating them with practical examples, the book significantly enhances accessibility for non-technical readers while also reinforcing its real-world applicability. This is an outstanding and highly valuable guide for small businesses navigating the path to sustainability.
Koray Kotan, Managing Partner, Ko-Solar

As weather extremes and natural disasters occur with increasing frequency due to climate change, a guide to help small businesses respond could not come at a better time. Kudos to Elvin for taking note of new realities including AI, virtual reality, and blockchain and her positive message: Businesses that act boldly will thrive, cutting costs, improving efficiency, and meeting customer expectations.
Alan Miller, co-author of *Cut Super Climate Pollutants Now!*

Resetting Our Future: Navigating Clean Energy Choices

A CLEAR Guide for Small Business Decision-Makers

Resetting Our Future: Navigating Clean Energy Choices

A CLEAR Guide for Small Business Decision-Makers

Elvin Yüzügüllü

CHANGEMAKERS
BOOKS

London, UK
Washington, DC, USA

CollectiveInk

First published by Changemakers Books, 2026
Changemakers Books is an imprint of Collective Ink Ltd.,
Unit 11, Shepperton House, 89 Shepperton Road, London, N1 3DF
office@collectiveinkbooks.com
www.collectiveinkbooks.com
www.changemakers-books.com

For distributor details and how to order please visit the 'Ordering' section on our website.

ISBN: 978 1 917704 01 4
978 1 917991 02 5 (ebook)
Library of Congress Control Number: 2025939647

A CIP catalogue record for this book is available from the British Library.

Design: Lapiz Digital Services

UK: Printed and bound by CPI Group (UK) Ltd, Croydon, CR0 4YY
Printed in North America by CPI GPS partners

The manufacturer's authorised representative in the EU for product safety is:
eucomply OÜ - Pärnu mnt 139b-14, 11317 Tallinn, Estonia,
hello@ eucompliancepartner.com, www.eucompliancepartner.com>

We operate a distinctive and ethical publishing philosophy in all areas of our business, from our global network of authors to production and worldwide distribution.

The ***Resetting Our Future*** Series

At this critical moment of history, with a pandemic raging, we have the rare opportunity for a Great Reset – to choose a different future. This series provides a platform for pragmatic thought leaders to share their vision for change based on their deep expertise. For communities and nations struggling to cope with the crisis, these books will provide a burst of hope and energy to help us take the first difficult steps towards a better future.

—Tim Ward, publisher, Changemakers Books

What If Solving the Climate Crisis Is Simple?

Tom Bowman, President of Bowman Change, Inc., and writing-team lead for the U.S. ACE National Strategic Planning Framework

Zero Waste Living, the 80/20 Way

The Busy Person's Guide to a Lighter Footprint

Stephanie Miller, Founder of Zero Waste in DC, and former Director, IFC Climate Business Department

A Chicken Can't Lay a Duck Egg

How COVID-19 Can Solve the Climate Crisis

Graeme Maxton, (former Secretary-General of the Club of Rome), and Bernice Maxton-Lee (former Director, Jane Goodall Institute)

A Global Playbook for the Next Pandemic

Anne Kabagambe, former World Bank Executive Director

Power Switch

How We Can Reverse Extreme Inequality

Paul O'Brien, Executive Director, Amnesty International USA

Impact ED

How Community College Entrepreneurship Creates Equity and Prosperity

Rebecca Corbin (President & CEO, National Association of Community College Entrepreneurship), Andrew Gold and Mary Beth Kerly (both business faculty, Hillsborough Community College)

Empowering Climate Action in the United States

Tom Bowman (President of Bowman Change, Inc.) and Deb Morrison (Learning Scientist, University of Washington)

Learning from Tomorrow

Using Strategic Foresight to Prepare for the Next Big Disruption

Bart Édes, former North American Representative, Asian Development Bank

Cut Super Climate Pollutants, Now!

The Ozone Treaty's Urgent Lessons for Speeding Up Climate Action

Alan Miller (former World Bank representative for global climate negotiations), Durwood Zaelke (President and founder, the Institute for Governance & Sustainable Development) and Stephen O. Andersen (former Director of Strategic Climate Projects at the Environmental Protection Agency)

Resetting Our Future: Long Haul COVID: A Survivor's Guide

Transform Your Pain & Find Your Way Forward

Dr. Joseph J. Trunzo (Professor of Psychology and Department Chair at Bryant University), and Julie Luongo (author of *The Hard Way*)

SMART Futures for a Flourishing World
A Paradigm Shift for Achieving Global Sustainability
Dr. Claire Nelson, Chief Visionary Officer and Lead Futurist, The Futures Forum

Rebalance
How Women Lead, Parent, Partner and Thrive
Monica Brand, Lisa Neuberger & Wendy Teleki

Provocateurs not Philanthropists: Turning Good Intentions into Global Impact
Maiden R. Manzanal-Frank, President and CEO, GlobalStakes Consulting

Feeding Each Other: Shaping Change in Food Systems
Nicole Civita (Vice President of Strategic Initiatives at Sterling College, Ethics Transformation in Food Systems) and Michelle Auerbach (Educator and founder of Modaka Communications)

Unquenchable Thirst
How Water Rules the World and How Humans Rule Water
Luke Wilson and Alexandra Campbell-Ferrari (Co-Founders of the Center for Water Security and Cooperation)

Why American Baby Boomers Are Uniquely Responsible for the Climate Crisis – and What to Do About It.
Lawrence Schick MacDonald, former Vice President of Communications, World Resources Institute

www.ResettingOurFuture.com

Previous Title

Synergies for Sustainable Energy
(ISBN: 978-1-60807-543-0)

Contents

This book is dedicated to my loving parents, who have always encouraged my passions, supported me in all of my endeavors, and believed in me. Without you, I would not be the person I am today, nor would this book have taken shape.

Preface: Who Is This Book For?

Small businesses help local economies grow, create jobs, and support communities. They quickly adapt to market changes with fresh ideas and new products. Many now focus on clean energy and sustainability to save money, stay strong, meet customer demands, and follow government rules. But business leaders face challenges, such as limited funds, lack of technical know-how, and fear of risk. These hurdles make it hard to pick the right clean energy options and upgrade systems in a fast-changing market. This book introduces a simple, five-step approach to help small business leaders make smart clean energy choices that fit their business needs.

Climate change and environmental damage create both problems and possibilities for businesses of all sizes. Large companies often get blamed for high carbon emissions, but smaller firms in their supply chains – those providing goods and services – cause over 80% of these emissions.[1] The World Economic Forum says that eight major supply chains, responsible for most global emissions, could cut them by 40% with simple, low-cost changes.[2] Reducing emissions from small businesses is key to reaching global clean energy goals.

Small business leaders who ignore carbon emissions risk losing ground to competitors and facing financial and environmental problems. Wasting energy and resources hurts profits and weakens businesses. Investment firms now check a company's climate impact before funding them, so ignoring emissions can limit financial options. As extreme weather grows more common, insurers may also refuse coverage, adding financial stress for businesses trying to recover.

The biggest risk of climate change to small businesses is direct damage, such as storms disrupting operations or supply shortages from dwindling resources. As the world moves toward sustainability, businesses will also face new technology, market, and policy changes that may raise costs or weaken the value of their products and services. Beyond physical and financial risks, companies will feel growing pressure from rules and customer expectations. Governments may demand cleaner energy use, and customers increasingly favor eco-friendly businesses.

Small businesses, unlike large firms, have fewer resources and a lower tolerance for failure, making them more cautious. They need funding to invest in clean energy and efficiency, but over time, the savings will be greater than the costs. They will also be better prepared for the effects of climate change. Clean energy can lower costs, improve products and services, drive new ideas, open new markets, boost competitiveness, and help small businesses qualify for green financing.

Are you a small business leader looking for a CLEAR way to make clean energy decisions? Read on…

Acronyms

AI artificial intelligence
AR augmented reality
BEV battery electric vehicle
CAES compressed air energy storage
CHP combined heat and power
CNG compressed natural gas
CSP concentrating solar power
DME dimethyl ether
EREV extended-range electric vehicle
EU European Union
FCEV fuel cell electric vehicle
GHG greenhouse gas
HEV hybrid electric vehicle
ICE internal combustion engine
LaaS lighting-as-a-service
Li-ion lithium-ion
LNG liquefied natural gas
LPG liquefied petroleum gas
MCDM multi-criteria decision-making
mpg miles per gallon
NaS sodium sulfur
OTEC ocean thermal energy conversion
PHEV plug-in hybrid electric vehicle
PV photovoltaics
RNG renewable natural gas
SMES superconducting magnetic energy storage
U.S. United States
VR virtual reality

Chapter 1

Why Care, and Why Now?

Remember when computers took up an entire room? It really wasn't that long ago. And just one process would take an eternity to complete. Then came the day when we could fit it all on top of a desk and perform multiple operations at the same time. Even that appears amusing now, as we hold what appears to be the entire world in our hands – email, social media, podcasts, web search, you name it – all accessed simultaneously and within seconds from the palms of our hands via cell phones or even smart watches.

How about those fat tube televisions? Didn't even have color in the beginning, and only one or two channel options. Then came color and a few more channels. Shortly after we thought that was great, along came hundreds of channel options and much slimmer devices approaching cinematic size. Now we can even customize when and where we view our favorite shows, thanks to streaming from multiple devices and not just our televisions.

These form and function changes happened fast, in a matter of decades, and during our lifetimes. For many consumer goods and in most economic sectors, we now have faster, more efficient, and more user-friendly technologies thanks to rapid and significant changes in size, form, and function.

What about the energy industry? Why do we still see the traditional shape and motion of oil platforms? Why are we still surrounded by smokestacks spewing pollutants and greenhouse gases (GHGs) into the atmosphere? Why is our energy system so "fossilized"?

Why should we care about transitioning to renewable energy, especially now?

Why? Because of the Challenges

Energy powers almost everything we rely on, but sustainable development has lagged. While clean energy has grown, it still supplies only a small share of global demand – far less than necessary. Despite increased awareness of environmental risks, many obstacles slow progress toward sustainability. Uncertainty, social norms, vested interests, and human resistance make change difficult. Time is running out. We've lost the luxury of gradual decisions. The longer we delay, the harder it is to undo damage and find timely solutions.

Climate Crisis and Air Pollution

Climate change worsens natural disasters, conflicts, and health crises. The cost of building a clean energy system is minor compared to the consequences of inaction. Moreover, fossil fuel combustion releases harmful pollutants, affecting air quality, human health, and ecosystems. Acid rain and smog harm wildlife and crops.

Energy Supply and Demand

Fossil fuels are becoming less viable due to shrinking reserves and environmental impacts. This creates risks, especially in fast-growing economies such as China and India. The transportation sector faces rising costs as oil supplies tighten. Major natural gas producers now rely on riskier extraction methods. Despite China's investment in nuclear energy and renewables, it remains the largest coal consumer. Nuclear power, while carbon-free, presents challenges in cost, safety, and waste disposal. Advanced modular reactors may offer more flexible solutions.

Natural Resource Geopolitics

Russia's 2022 invasion of Ukraine shook global energy markets. Russia used its fossil fuel exports to pressure dependent nations, especially in the European Union (EU). In response, the EU plans to cut Russian imports by 2027,[3] replacing them with liquefied natural gas from the United States (U.S.)[4] and expanding renewables. But these shifts bring immediate risks, including potential energy shortages. Some EU nations may reconsider coal use.[5] Russia may seek new markets, such as China, possibly slowing global decarbonization efforts.

Energy-Water-Food Nexus

Water, food, and energy are deeply linked. Water is needed for growing food and generating energy. Energy powers water systems and farming. As the global population grows, pressure on these systems increases, especially in water-scarce regions with high populations and farming activity. Climate change makes these challenges worse – extreme weather damages infrastructure and drains freshwater supplies. This forces the use of energy-heavy solutions such as desalination and advanced pumping, raising energy demand.

Global Health and Climate Change

The COVID-19 pandemic disrupted health and economies, hitting vulnerable groups hardest. It showed how vital energy is in health care. Limited energy access slowed hygiene efforts and increased indoor air pollution risks. Reliable energy is essential for running medical equipment, sterilizing tools, and storing medicines. Strong energy systems mean better health care.

Why? Because of the Opportunities

As the population grows and energy demand rises, we face two major challenges: reducing carbon emissions and building a stable, affordable energy system. Climate change makes action urgent. Right now, we have the chance to create real change. With a variety of advanced clean technologies and centuries of scientific progress, we can develop sustainable energy solutions – if we act quickly and think creatively.

Being Safe Is Risky

We often stick to familiar ways of doing business because they feel safe, even when they carry hidden risks. Resistance to change holds many back. But today, businesses worldwide are realizing the environmental and economic advantages of clean energy and sustainable practices. Regulations are also shifting, such as the Securities and Exchange Commission's 2022 proposal requiring businesses to disclose climate risks and greenhouse gas emissions. There is no universal fix for our energy and environmental challenges. Instead, we need a mix of solutions working at the same time. Clean energy technologies now allow businesses to tailor approaches based on local resources and specific needs. By combining different technologies, industries can solve multiple problems at once, reduce risks, and create new market opportunities.

Falling Costs and Rising Investments in Renewables

Renewable energy has become much more affordable over the past decade, now competing with fossil fuel generation costs. Between 2010 and 2020, global costs for key renewables dropped significantly: solar photovoltaics (PV) by 85%, concentrating solar power by 68%, onshore wind by 56%, and offshore wind by 48%.[6] This shift has made renewables some of the cheapest sources of new energy, a dramatic change from when they were more than twice as expensive as the priciest fossil fuels.

Investor interest is also changing. More funds are shifting away from fossil fuels and toward clean energy, signaling strong confidence in sustainable solutions. Between May 2016 and May 2021, the S&P Clean Energy Index grew by 22%, while the fossil-fuel-heavy S&P 500 Energy Index fell by 4%.[7] Investments in clean energy technologies are increasingly seen as an inevitable shift, rather than just a response to climate change.

Energy Access and Resilience

Every country, region, and community needs reliable energy to grow and strengthen its economy. Whether from local sources or international markets, access to affordable energy is essential. Fossil fuels are unevenly distributed and rapidly depleting. Renewables, however, vary in abundance but are available everywhere. Every region gets some sunlight, and most have wind resources. Waste-to-energy solutions are also becoming more viable, as waste is a constant part of human activity.

Green Jobs

In 2023, clean energy jobs made up 56% of new energy sector employment in the U.S., growing at twice the rate of the rest of the energy sector and the broader economy. Clean energy also accounted for 42% of total energy jobs and nearly 5% of all new jobs nationwide.[8]

Globally, in 2022, there were 13.7 million renewable energy jobs, with solar PV leading employment. Nearly two-thirds of these jobs were in Asia, with China making up 41% of the total.[9] The shift to renewables is expected to create more jobs than it replaces in fossil fuels. Many workers in fossil fuel industries can transition to clean energy through retraining, carrying over their expertise.

For small businesses, embracing clean technologies can lower costs, boost resilience, and ensure competitiveness in a changing market. With all the risks and opportunities in mind, let’s explore a practical five-pillar strategy to navigate clean energy and make informed decisions for a sustainable future.

Chapter 2

How Do You Navigate? The CLEAR Approach

A Different Perspective

We often solve problems by analyzing them, essentially breaking them into smaller parts and studying each piece separately. While this approach helps in many cases, it can cause us to focus too much on details and miss how different parts connect. This can lead to new and harder problems later. While analyzing issues has value, complex decisions need a broader view – **systems thinking.**

A system is simply a set of connected parts working together as a whole. A systems thinker sees how all parts link and understands that the whole is more than just its individual pieces combined. This perspective helps us grasp both the system and its parts at the same time, as well as how they interact with each other and the outside world. Systems thinking helps us recognize patterns instead of isolated snapshots and reduces the risk of unexpected problems when making changes.

The CLEAR approach and its five pillars use systems thinking to support better, more complete decision-making. Now let's explore how.

The Blind Men and the Elephant

This parable, rooted in Buddhism but found in many texts, warns against closed-mindedness and perfectly illustrates the value of systems thinking.

There was once a cruel king, who one day ordered that a group of men from his kingdom – those who were blind from birth – be gathered and presented with an elephant. He then asked them what they thought of this powerful animal. The blind men each touched different parts of the animal. One touched the trunk and said, "This being is like a thick snake." Another touched a leg and thought it was like the trunk of a tree. "No," said another, who had reached an ear, "this is like a fan!" And so it went on: "like a wall" (the body), "like a rope" (the tail), and "just as smooth and hard as a spear" (the tusk).

The Buddha teaches that our perspective is limited, meaning we never see the full truth. Like the blind men, if we focus on isolated parts and ignore connections, we fail to grasp how a system truly works.

The CLEAR Approach

Do you feel like your actions and decisions won't really move the needle? Or do you continue with your business-as-usual approach, denying problems and thinking everything will be just fine? You may put in a fair struggle and make some incremental improvements. But none of these approaches really gets you closer to the most beneficial decisions or outcomes.

Every energy-related decision on the path toward sustainability has many layers of complexity and may often feel too overwhelming to tackle. Information is abundant but varied and not necessarily well-organized. You want to make the best decision for your small business, your employees, and the planet, but don't know where to start.

So then, what's a better solution?

A **CLEAR** framework can simplify these decisions and provide a robust foundation for moving forward. The five pillars of the CLEAR framework are to:

1. **(C)onsider:** Shift your perspective and change your mindset.
2. **(L)essen:** Start with efficiency to reduce demand and wasted energy.
3. **(E)xtend:** Use circular economy principles to help extend the life of existing materials and products and thus help reduce carbon.
4. **(A)ssemble:** Use a portfolio of technologies in an integrated way to leverage synergies.
5. **(R)elate:** Involve all stakeholders and their priorities to make better decisions.

Every energy decision involves technical, economic, and practical factors. Before focusing on implementation, use the CLEAR framework as a foundation and checklist to ensure the best decisions for your business and the environment. Now, let's examine each component of the CLEAR approach.

CONSIDER: Shift Your Perspective and Change Your Mindset

This pillar **anchors** the CLEAR approach, guiding how we evaluate and apply the other four pillars. So, what does shifting your perspective mean?

Before jumping into technologies and processes to solve energy challenges, we must first step back and think strategically. Acting too quickly on big issues often leads to weak solutions or more difficult paths. At the core of this approach is a mindset shift – not just in the planning stage but throughout the entire lifecycle of the solution. This shift means redefining how we approach energy problems. Instead of relying on quick fixes, we take a broader view, incorporating environmental, economic, and technological factors from the start. By embedding strategic, systematic thinking early on, we create energy solutions that are not only effective from the beginning but also adaptable to future challenges.

And that starts with changing our frame of view. Problems often seem unsolvable inside a specific frame. But if we create a new frame or enlarge the current frame, problems tend to vanish, and new opportunities have a chance to appear. Yes, the closer a solution is to the current way of doing things and the more familiar the customer experience, the easier the transition and technology adoption will be. But what if we expanded those boundaries in our thinking to figure out what other options we may have? It's a new era, the technologies are new, and these are big changes that will define life for generations. So, we should be more thoughtful, strategic, and flexible in our thinking.

Consider our current transportation sector. It is built for gasoline-fueled vehicles, but alternative fuels such as hydrogen and electricity require different infrastructure. For example, hydrogen can come from various sources, requiring careful planning for its production, storage, and delivery. And electric cars take longer to charge than gasoline cars do to refuel. So why are we forcing new energy solutions into outdated systems? We need to rethink infrastructure. Big-box store parking lots could serve as electric vehicle charging hubs, allowing cars to charge while people shop. And what if carports at stores and resting stops integrated solar PV, providing both shade and renewable electricity? These innovative approaches, already in development, show how we can use our creative thinking to design better systems.

Consider a factory as another example. Manufacturing processes can generate significant air pollutants, toxic water, and waste. Instead of managing harmful emissions after they occur, we could rethink the process itself. By modifying steps to *prevent* or *reduce* pollution, we make control easier. Adjusting operation timing could also tap into curtailed wind energy, which is often wasted due to transmission limits or excess generation.

Transitioning to a renewables-based economy without foresight can create new challenges. Fossil fuels once seemed efficient and harmless, but we now see their long-term downsides. Today, vested interests, inertia, and short-term thinking blur our judgment. The good news? Our accumulated experience offers insight into these challenges. And even with full renewable energy adoption, we must still plan for their eventual environmental impact. Otherwise, we will end up with a major waste management issue. Solar panels last 25–30 years but eventually degrade and need dismantling. As technology advances, older systems will need replacement. Critical minerals for clean energy technologies are finite and geographically limited. Without strategies for reducing reliance, reusing materials, and improving recycling, we risk resource shortages and waste problems.

To make better decisions, *shift from reacting to thinking strategically*. Before acting, step back and avoid quick fixes. Consider environmental, economic, and technological factors. Design flexible solutions and expand your thinking beyond current systems to explore new approaches for future challenges.

LESSEN: Start with Efficiency to Reduce Demand and Wasted Energy

Consider:

- Where do we see the greatest energy losses in our processes and the best opportunities for conserving energy?
- How can we change our processes or behaviors to avoid wasting energy?
- What simple and cost-effective strategies can we start with to realize quick gains?

A significant amount of energy is lost across industries. So, before tackling energy challenges, we must first reduce wasted energy, improving **efficiency**. This may sound very basic and even insignificant, but it is a key first step. Businesses can take simple, low-cost steps to cut energy waste. For example, plug-and-process loads – devices plugged into outlets and hardwired equipment – account for approximately 47%[10] of primary energy use in commercial buildings in the United States. Powering down all electronics after business hours, consolidating loads, and replacing old appliances with more efficient models can prevent this waste. Similarly, industrial facilities can cut costs by shutting down unused energy-intensive machinery such as motors, with savings rising based on motor size and electricity rates. Regular maintenance of heating, ventilation and air-conditioning systems, compressors, and/or steam systems helps detect inefficiencies and leaks, which can be repaired and insulated to reduce waste and save money.

Electricity Generation

The U.S. electric grid, developed in the late nineteenth century, evolved from local independent utilities into a vast system where large, centralized power plants send electricity over long distances via transmission and distribution lines. This setup results in about 60%[11] of the primary energy being lost before reaching consumers, mostly during power generation. Fossil fuel plants operate at just 30% to 40%[12] efficiency. And as electricity travels to consumers, about 5%[13] is lost due to electrical resistance in wires and voltage changes in transformers, despite high-voltage designs to minimize these losses. Additional losses occur when electricity is converted into usable energy. Expanding fossil fuel use requires more power plants and infrastructure, increasing costs for consumers. Distributed generation improves efficiency by producing electricity closer to where it's used, reducing energy losses. Solutions range from home or business solar PV systems to microgrids (explained further in the following chapter) for industrial sites or military bases. As climate-driven extreme weather intensifies, distributed generation can strengthen grid stability and security by providing reliable backup power during outages.

Industrial Processes

We don't often think about how things are made, but every product we use – phones, clothes, paper, appliances, cars, and more – is made in factories with complex processes that consume energy and natural resources. Before production, upstream activities such as feedstock extraction and transportation also rely on fossil resources, making the entire supply chain heavily dependent on them. The industrial sector consumes about one-third of total energy in the U.S., and about half of total energy globally.[14] About 20% to 40% of energy is lost in heavy equipment like boilers, pumps, and fans, and up to 15% more via motors and process heat.[15] Quite often, things like failed steam traps, poor combustion control, and leaking compressed air systems waste significant energy. Leaks force equipment to work harder for the same useful output – the very definition of being inefficient. Aging, vibration from heavy machinery, chemical exposure, or improper fitting make industrial components prone to defects, with energy losses often unnoticed until they prove costly. For example, about 80% of air leaks from compressors are not audible, and by the time they do become audible, a simple quarter-inch leak could waste more than $17,000 a year. In fact, poorly maintained compressed air systems in the U.S. add up to $3.2 billion per year in wasted utility payments.[16]

While industrial energy waste can be significant, simple fixes can offer the greatest opportunity for energy savings. Prioritizing regular audits and proactive maintenance practices is essential, as it goes beyond ensuring safe and regular operations; it is the first step in recovering wasted energy and dollars. So, before jumping to meet a large energy demand at a manufacturing facility, one can simply start by looking at energy **wasted** throughout the process. It makes little sense to buy more equipment to keep up, perpetuating the high energy use while also adding to the financial burden of maintaining another asset. It's like trying to drain a boat that has a hole: no matter how many buckets of water you dump, more water will keep coming in through the hole left unplugged.

Waste heat in industrial settings can be captured and reused, reducing costs and emissions. Vast amounts of heat are lost from equipment exhaust gases, hot product streams (e.g. hot cast steel), and even hot equipment surfaces (such as aluminum sidewalls). Recovered heat can be converted into electricity, used directly in industrial processes (e.g. pre-heating, drying), or used in applications such as heating water, cooling (via absorption coolers that use heat energy instead of electrically driven compression), and space heating.

Transportation

For over a century, cars have been running on one fuel (gasoline) and one technology (internal combustion engines). In the late 1800s, steam and electric vehicles were more common. Steam had been used in trains and boats, and its use in cars was safe but cumbersome and slow. These cars avoided the strenuous hand-cranking of gasoline engines and had longer ranges than early electric cars, which, despite being quiet and emission-free, were limited by battery life. The launch of Henry Ford's mass-produced Model T in 1908, combined with innovations such as the electric starter, quickly popularized gasoline cars. These cars were more efficient than steam cars and were supported by abundant petroleum. However, with millions of cars on the road today and fossil fuel resources declining, the landscape has changed.

Car efficiency has improved, but incremental gains are becoming less effective, making a shift to new technologies necessary. For example, switching from a 10 miles-per-gallon (mpg) sports utility vehicle to a 15 mpg one would save you more gallons of fuel (traveling the same distance) than switching from a 30 mpg compact car to one that gets 35 mpg. Both scenarios achieve a 5-mpg improvement, but as efficiency increases, further gains become harder to achieve.

Incremental efficiency improvements are no longer enough – we need new technologies and platforms for greater energy savings. The transportation industry is shifting to an electricity-based system that supports multiple technologies such as full battery, plug-in, or hydrogen fuel cell vehicles, offering more than just efficiency gains. This shift diversifies fuel sources, reducing reliance on a single energy type. Moreover, transportation and electricity sectors are also merging. For example, plug-in vehicles not only draw power from the grid but can also serve as backup storage for the electric grid.

Chart Your Path: Energy waste leads to an inefficient business. Before adopting new solutions, always prioritize improving energy efficiency. Reducing waste is crucial and impactful, given the significant energy losses across sectors. Simple, low-cost actions can yield substantial savings and make broader energy challenges more manageable.

EXTEND: Use Circular Economy Principles to Help Extend the Life of Existing Materials and Products and Thus Help Reduce Carbon

Consider:

- How do our materials and water use impact our energy use, and vice versa?
- What processes can we change so that we generate less or less harmful waste and keep materials and products in use?

- Can we offer some of our goods as a service?
- Can we share resources with adjacent industries?

We have been on a linear "take-make-waste" path which relies heavily on using fossil fuels. A **circular economy**[17] provides a systems-level perspective and is driven by innovative, design-based thinking. It is built on three main principles:

1. **Eliminating Waste and Pollution:** Reducing waste means that we also reduce the emissions tied to the production of materials and products.
2. **Keeping Materials and Products in Use:** Extending the life of products preserves embodied energy – the energy used to process, manufacture, transport, and use materials – rather than creating new ones that increase greenhouse gas (GHG) emissions. In a linear economy, discarded products lose the opportunity to have valuable byproducts reclaimed. Circular economy strategies combat this by reusing, refurbishing, remanufacturing, and recycling, conserving resources[18] compared to virgin material use.
3. **Regenerating Natural Systems:** By regenerating our natural systems, carbon is kept in the soil, helping build soil health.

A key idea in the circular economy is shifting from selling goods to offering them as a service: a subscription model. For instance, Signify (formerly Philips Lighting) provides "lighting-as-a-service" (LaaS).[19] It retains ownership of the lighting systems while handling installation, maintenance, and upgrades. Customers pay a monthly fee instead of upfront costs.

This model encourages Signify to produce durable, easily recyclable or reusable products, cutting waste and resource use. At Amsterdam Schiphol Airport, LaaS reduced energy use by 50% compared to the previous system.

Electricity and transportation account for 55% of global GHG emissions, which can be reduced through efficiency and clean technology. The remaining 45% – from industry, agriculture, and land use – is harder to address.[20] While innovation and carbon capture help, circular economy strategies are key to fully tackling these emissions. Only 9% of the world is circular today, so there is a significant potential for positive change.[21]

By 2050, global demand for steel, cement, aluminum, and plastics could double or quadruple, while food demand may rise 42%. Despite efficiency gains and zero-carbon technologies, GHG emissions from just these four materials could reach 649 billion tons by 2100, exceeding the 420-580 billion-ton carbon budget for the 1.5°C climate target. Circular economy practices, such as recycling, reusing, reducing waste, eliminating food waste, and regenerative agriculture, could cut emissions by 40% in industry and 49% in food systems by 2050.[22]

A circular approach enhances climate resilience by keeping materials in use, reducing reliance on raw materials vulnerable to climate risks. Regenerative agriculture improves soil health, and healthy soil improves water absorption and retention, enhancing resilience to extreme drought and heavy rainfall.

Chart Your Path: Embrace circular economy principles to extend material life and cut carbon emissions. Prioritize keeping products and materials in use through maintenance, reuse, refurbishment, remanufacture, recycling, and composting. By offering goods as a service or sharing resources with other industries, you can lower emissions, reduce costs, and boost climate resilience.

ASSEMBLE: Use a Portfolio of Technologies in an Integrated Way to Leverage Synergies

Consider:

- Are we considering multiple technologies and resources, or relying on a single approach that could create vulnerabilities?
- How can we leverage the synergies of different technologies to maximize efficiency, sustainability, and adaptability to future challenges?
- What local resources and demands should influence the optimal mix of technologies for our specific needs?
- Are we exploring ways to extract additional value from single resources, to reduce waste and increase profitability?

We're at an exciting point in history with a wide portfolio of clean technologies that are feasible, cost-effective, and impactful. The next chapter explores these innovative technologies and approaches. Why not leverage the strengths and benefits of several technologies to create the optimal system? We no longer must rely on singular technologies, resources, or approaches for our energy needs. Yes, we're on a compressed timeline, and there are deeply ingrained interests and many layers of complications. But we don't have to tackle these difficulties with limited tools. We not only have a wide range of technology options, but many of them are rapidly becoming cost competitive with established pathways.

And just like in investing or many other areas of life, why put all your eggs in one basket and create vulnerabilities? Especially considering the dynamics and risks of climate change, we cannot afford to rely on just one technology or pathway. There is not going to be one silver bullet that magically solves all our problems. Each sector, subsector, region of the world, and so on, is going to have specific technologies or sets of technologies and pathways that best meet its needs. By strategically combining technologies for efficiency, energy production, and storage, we can optimize outcomes based on available resources and demand. Microgrids are a great example of this strategy. They integrate renewable and, at times, non-renewable sources to supply heat, electricity, and water for communities or facilities.

This approach can optimize a single resource by extracting additional value beyond its primary use. For example, emerging biorefineries are analogous to how today's petroleum refineries take in a single feedstock (crude oil) and convert it to a range of petroleum products that serve multiple needs. The input here is a single biomass feedstock, but through processing into multiple intermediaries, a spectrum of value-added products (food, feed, chemicals, and materials) and bioenergy (biofuels, power, and/or heat) can be created. The goal of this integration is to maximize biomass feedstock value by continuously producing bioenergy and valuable bioproducts while recycling waste with minimal disposal. This benefits both the environment and the bottom line. Facilities can vary, producing high-value, low-volume chemicals alongside low-value, high-volume fuels. This approach enhances energy flexibility, lowers costs, reduces emissions, and boosts profitability. And bio-based materials such as biochar improve soil organic carbon, supporting agriculture.

One widely recognized approach to breaking down a big problem, instead of chasing a magic, silver bullet solution, is the **stabilization wedges** concept created by Robert Socolow and Stephen Pacala of Princeton University.[23] This concept puts forward the idea that seemingly insurmountable carbon emissions can be tackled by dividing the challenge into **wedges** of more manageable portions. The difference between the current emissions trajectory and the expected increase forms a triangle, called the **stabilization triangle**, which is then divided into equal pieces (or wedges). Each wedge represents a strategy to cut emissions, such as wind power, industrial efficiency, carbon capture, and so on. This approach spreads the burden across multiple solutions, making emissions reduction more achievable.

Chart Your Path: Don't rely on a single technology or approach to meet your energy demands. Diversify solutions to enhance resilience, efficiency, and sustainability. Strategically integrate multiple technologies and approaches to navigate climate and market changes, leveraging each technology's strengths for overall economic benefit.

RELATE: Involve All Stakeholders and Their Priorities to Make Better Decisions

Consider:

- Have we considered the perspectives and priorities of all stakeholders involved in or impacted by our decisions?

- Are we allowing subjective biases or past experiences to overly influence this decision, or have we analyzed all alternatives thoroughly?
- What criteria are the most important for the technology and/or process alternatives we are considering?
- How can we integrate both qualitative and quantitative factors to ensure a balanced and transparent decision process?

When choosing energy technology, it's crucial to consider multiple perspectives, tradeoffs, and stakeholder input. Multi-criteria decision-making (MCDM) structures subjective judgments, leading to more transparent and effective decisions. Given the complex nature of real-world judgments, this framework allows decision-makers to weigh both quantitative and qualitative criteria simultaneously.

- **Step 1:** Structure the decision problem into a *hierarchy*, placing the main objective at the top, followed by criteria and sub-criteria, with alternatives listed below for evaluation.
- **Step 2:** *Compare criteria* one at a time by asking: "How much more important is criterion A than criterion B?" Use a nine-point scale to measure preference strength.
- **Step 3:** Use preference ratios from step 2 to *assign weights*, then normalize them so their total equals one.
- **Step 4:** *Compare alternatives* in pairs for each criterion by asking: "With respect to criterion A, how much more do you prefer alternative X to alternative Y?" To support these comparisons,

gather performance data through literature reviews, expert input, or detailed analyses (e.g. solar vs. wind power lifecycle emissions).

- **Step 5:** Multiply each alternative's score (from step 4) by the weighted criteria, sum the results, and *rank the alternatives* based on total scores.

MCDM can help guide energy sector decisions, including technology investments, local planning, building energy performance, and selecting energy technologies and approaches.

Example Application:[24] The analytical hierarchy process helped select energy-efficient public lighting options for Brazilian cities. To begin, global best practices for energy-efficient public lighting in cities were categorized into four programs: audit, retrofit, alternative energy, and management. The main objective was to identify the best lighting practice for each municipality. The criteria considered under this objective were costs, energy savings, and ease of implementation. Public lighting experts completed an online questionnaire, comparing criteria and alternatives in pairs. Their responses were processed in decision-making software to generate final judgments. The selected alternatives could achieve up to 6 GWh/year in energy savings and $600,000/year in financial savings.

Example Application:[25] A Toyota plant in Ovar, Portugal, sought to optimize its operations, focusing on its painting sector, which consumed 70% of total costs and was the primary bottleneck in production. Using multi-criteria decision analysis, the team evaluated eight painting plans (PP-A to PP-H) based on key criteria of quality (60.6%), energy use (23%), paint consumption (12.6%), and vehicle output (3.9%). Pairwise comparisons of historical data (June-December 2012) determined weights. Through this structured evaluation, PP-C, featuring simple vehicles with three colors, emerged as the best option, improving cost, energy efficiency, and quality. This plan significantly contributed to painting process optimization by balancing cost, energy efficiency, and production quality. The approach encouraged in-depth discussions, refining key performance indicators and setting clear targets for the painting sector. Data-driven insights helped identify efficiency improvements, reducing costs and streamlining operations.

Advancements in technology are enhancing decision-making efficiency. Artificial intelligence algorithms now automate data analysis, identify patterns, and improve predictions in MCDM models. Real-time data allows decision-makers to refine evaluations and adapt to changing conditions for better outcomes.

Chart Your Path: For complex energy decisions, use structured tools such as MCDM to balance tradeoffs. Define objectives, prioritize criteria, and engage stakeholders to evaluate alternatives systematically.

Chapter 3

What Technology Options Do You Have?

The clean energy landscape is far more diverse and richer in options than it initially appears. While technologies such as solar photovoltaics (PV), wind turbines, and batteries usually get the media spotlight, there are many more options to choose from. Each technology may be in a different stage of development, but the good news is that clean energy costs are decreasing (even approaching parity with conventional sources), different technologies can meet different needs, and integrating technologies will provide even more benefits than a single pathway.

This chapter showcases efficiency measures for every sector of our economy, a wide range of clean energy pathways, and diverse energy storage mechanisms. Then, it provides an overview of cutting-edge methods and technologies that leverage information technology, the natural world, and fresh perspectives on infrastructure design to address challenging problems. When putting the CLEAR approach into practice, look to this chapter for inspiration and a portfolio of options to choose from.

Energy Efficiency

Energy efficiency is the perfect place to start. It's a **low-hanging fruit**, offering a low-cost solution to growing energy demand. Being efficient means getting more value out of energy usage, by focusing on reducing **wasted** energy, while still enjoying the same comforts and benefits. Investing in energy supply is significantly more expensive than investing in efficiency. While alternative energy technology research, development, and deployment can take years, energy efficiency initiatives can be implemented and benefits realized quickly, typically within a couple of years, or even a few months. This enables businesses to reduce costs or gain a competitive advantage. By increasing the efficiency of our homes, cars, and industries, we can use fewer fossil fuels (lowering harmful emissions), postpone costly grid capacity upgrades, and protect ourselves from energy supply and price disruptions. But efficiency alone won't lead to sustainability. Rather, it should be included in a portfolio of goals. And alternative energy technologies must continue to be developed and implemented.

Efficient Buildings

Buildings mainly use energy for space heating and cooling, water heating, and lighting. Depending on the specific function of the building, these energy needs might differ somewhat. Office buildings, for example, consume more energy for lighting and have an additional heat load from equipment and humans; hotels and resorts require a lot of water heating; and refrigeration energy is important in the food industry. Buildings that are designed and retrofitted with better windows, more insulation, and other features help manage heat losses and thus reduce energy demand. Moreover, commercial benchmarking laws have been enacted in several markets in the United States, where regulations require building owners to disclose information about their property's energy consumption. As tenants are responsible for paying energy-related costs, building owners have traditionally had little incentive to invest in energy-saving measures, but now building owners are also incentivized to invest in these efficiency improvements.

Efficient Vehicles

Fuel economy refers to how far a vehicle can travel on a unit of fuel, measured in miles per gallon (mpg) or liters per kilometer. The Corporate Average Fuel Economy standards, established by the U.S. Congress in 1975, mandate that automakers achieve fleet-wide average fuel economies. In June 2024, the National Highway Traffic Safety Administration set new standards, targeting an average fuel economy of 50.4 mpg by 2031 for light-duty vehicles and 35 mpg by 2035 for heavy-duty trucks and vans.[26] And as alternative fuels and vehicles become more common, consumers need clear information to make better decisions. Improvements to the vehicle fuel economy label provided by the U.S. Environmental Protection Agency and the National Highway Traffic Safety Administration now include detailed information such as driving range, charging times for electric vehicles, fuel or electricity usage per 100 miles, emissions ratings, and potential fuel savings over five years.

Efficient Industry

In the process of mining materials and manufacturing a wide range of products, the industrial sector consumes a significant amount of energy. Among the most energy-intensive industries are chemicals, petroleum and coal products, paper, and primary metals.[27] Industrial facilities are businesses with a lot of expenses and the goal of turning a profit. As a result, energy is yet another cost (albeit a significant one) to be reduced. Efficiency measures reduce energy consumption, reducing related costs and shielding industries from the risks and fluctuations associated with the high and variable costs of fossil fuels.

Energy is lost at various stages of manufacturing processes due to equipment inefficiencies, mechanical and thermal limitations, and poor maintenance. As a result, even minor efficiency improvements can result in significant energy savings and a reduction in industry's carbon footprint. Energy efficiency practices also reduce waste, which, if left unchecked, can have a negative impact on a company's bottom line. Moreover, managing energy consumption and costs provides a competitive advantage that attracts support from investors and benefits customers.

Combined heat and power (CHP) systems offer industrial facilities an efficient way to reduce energy use by simultaneously generating electricity and thermal energy from a single fuel source onsite at the facility. Unlike separate heat and power generation, which typically achieves 50% efficiency, CHP systems can reach efficiencies between 65% and 75%.[28] This process involves using a fuel – such as natural gas or biogas – to power a prime mover (steam turbine, gas turbine, fuel cell, etc.), generating electricity and capturing residual heat to produce steam. This steam can be used directly for industrial processes or to power a steam turbine for additional electricity. Moreover, when configured for trigeneration, CHP systems can also provide cooling by using waste heat (through absorption chiller technology) to produce chilled water for refrigeration or air conditioning.

Renewable Electricity

The core premise of renewable resources is that they are replenished at a faster rate than they are consumed. And the use of these resources in a variety of technologies does not emit any global warming-inducing greenhouse gases or toxic air pollutants. This isn't to say that using renewable energy has no consequences. Manufacturing solar panels or wind turbines, for example, still consumes resources and may create emissions during the manufacturing process. But the key point here is that with renewable energy, we not only do not have to worry about the resource itself running out, but renewables also have far fewer negative impacts, while their benefits far outweigh any drawbacks.

However, to avoid a waste problem from renewable energy technologies at the end of their useful lives, it's crucial to integrate recyclability and reusability into their design from the start. While recycling regulations and processes are important, addressing the issue **proactively during design** ensures that these technologies can be more easily, economically, and effectively recycled and reused later. By 2030, the U.S. is expected to decommission about 1 million metric tons of solar panels annually, with global figures potentially reaching 8 million metric tons. And by 2050, global waste from solar panels could hit 78 million metric tons, with lithium-ion batteries contributing another 11 million metric tons and wind turbine blades reaching 12 billion metric tons.[29]

Solar panels and wind turbines use some materials that can be harmful if not disposed of properly, such as lead and cadmium, but also use those that hold economic value if recycled, for example the rare metals tellurium and indium. As the first generation of these technologies reaches the end of their useful lives, the recycling challenges that arise are due to their complex designs and the variety of materials used.

For instance, solar panels contain materials ranging from aluminum and silicon to valuable metals, which need to be separated for recycling. Similarly, wind turbines include recyclable steel and copper, but their blades, made from composites like glass or carbon fiber and resin, are tougher to dismantle and recycle. Innovations such as Siemens Gamesa's turbine blades, which are manufactured using a new type of resin,[30] are making recycling more feasible by simplifying the separation process, enabling material recovery for new industrial uses. Solar panels, in addition to being recycled, can also be reused as-is or refurbished for applications like bike- and vehicle-charging stations or in remote locations off the grid, although this involves navigating various safety regulations.

The U.S. Inflation Reduction Act of August 2022 boosts domestic production of renewable energy technologies and promotes recycling of their components. It offers $10 billion in tax credits for manufacturers updating or building new facilities for renewable energy components, including a 30% credit for investments in recycling facilities. At the state level, California has facilitated the recycling of solar panels by reclassifying used panels as universal waste instead of hazardous waste, lowering recycling costs. New York is considering a similar policy. Illinois has established a task force to explore recycling options for renewable components, and Washington state mandates that manufacturers take back and recycle solar panels sold after July 1, 2017, starting from July 1, 2025. These initiatives reflect a broader commitment to the principles of the circular economy in managing renewable energy resources.[31]

Renewable energy technologies also have different economies than fossil-based technologies, because renewables typically have high initial investment costs, lower operating and maintenance costs, and very low or no fuel costs. Conventional fossil-based facilities, on the other hand, have lower initial investment costs but high (and in some cases dominant) fuel costs that are highly volatile. Renewable technologies are also more vulnerable to changes in capital costs and financing terms. And opponents of widespread renewable energy typically argue that there will always be a need for baseload electricity generation based on fossil fuels, as renewables are intermittent and thus not as dependable. However, as the following sections highlight, not only do we have a range of energy storage options to meet needs, but we also have other renewable resource options besides intermittent wind and solar; in addition, there are baseload renewable options such as geothermal energy.

Solar

Solar energy is a vast resource, with the amount of energy supplied by the sun in one day being sufficient to meet all the energy needs of the global population. The key here is to convert this vast resource into electricity as efficiently as possible. Solar energy can be used either passively or actively. By optimizing building design to make better use of daylight or heat, passive solar use collects energy from the sun without converting its heat or light into other forms. Active solar use involves converting the sun's energy into electricity using PV technologies (grid-connected or off-grid), converting the sun's heat into electricity using solar thermal technologies, or using the heat directly.

Solar cells are at the heart of a PV system. A module is formed by the combination of several of these cells, and it is the smallest PV component sold commercially. Several modules can also be combined to form an array for a higher total power output. There are three main categories of PV technologies on the market: monocrystalline, polycrystalline, and thin-film technologies. Monocrystalline and polycrystalline systems are based on silicon and dominate the market due to their low costs, high efficiency, and wide range of well-known manufacturers. Silicon is an abundant element in the Earth's crust and a semiconductor material that can be used in PV applications. PV systems based on thin-film technologies are still in the early stages of market adoption. Thin-film solar cells are composed of successive thin layers of solar cells deposited onto a large, low-cost substrate such as glass, polymer, or metal. These cells require significantly less semiconductor material than their crystalline counterparts to absorb the same amount of sunlight, resulting in lower costs but also lower efficiencies. Thin-film cells can also be integrated into building components such as windows to generate electricity for onsite use or grid export.

Solar cells made of perovskite, a mineral with a unique crystal structure, have gained popularity due to their high efficiency and low cost when compared to traditional silicon solar cells. Perovskite thin-film solar cells have rapidly increased in efficiency, from around 3% to 25% in a decade and a half, with potential to go much higher.[32] These cells can also be manufactured using simple and inexpensive processes, lowering production costs. Their lightweight and flexible nature makes them ideal for portable and wearable devices. They can also be made semi-transparent, allowing use as glazing in windows.

Solar PV panels do not need to always be placed on rooftops or large swaths of barren land. Situating them in areas such as canals, reservoirs, and agricultural land can serve the dual purpose of improving these spaces' functions while generating electricity. *Floatavoltaics*, or floating solar panels, can be mounted on top of bodies of water, reducing evaporation, which is beneficial in drought-prone areas, while producing clean electricity. The water also cools the panels, allowing them to operate more efficiently. Similarly, *agrivoltaics* integrates solar panels with farming practices, whereby panels and the crops share the land. In addition to providing an additional beneficial use for grazing land, PV panels help increase crop yields by providing shade for crops that require it and reducing the need for irrigation water. Another innovative approach, demonstrated by Ko-Solar, is to install solar panels on unused areas such as highway sound walls, railroad barriers, and airport tarmacs. This approach expands solar energy use without requiring more land. It also helps reduce noise and creates new revenue for local governments, airports, and rail operators.[33]

Concentrating solar power (CSP) technology uses mirrors or lenses to concentrate the sun's rays, which heats a fluid to produce steam. The rest of the process is like that of conventional power plants, in which steam drives a turbine and generates power. CSP plants focus the sun either along a line (parabolic troughs and linear Fresnel systems) or at a point (dish Stirling and solar tower systems). Power from parabolic trough, linear Fresnel, and solar tower systems can go up to 80–200 MW, whereas dish Stirling systems are deployed on a much smaller scale (3–25 kW)[34] and can be used in distributed power applications.

CSP technologies are generally limited to hot, dry regions because only strong direct sunlight can be concentrated to the temperatures required for electricity generation by these systems. But because these systems require a large amount of steam, water consumption can be significant, comparable to or even greater than that of traditional coal and natural gas plants. The regions with abundant solar resources that are suitable for CSP-based electricity generation are also often remote, separated by long distances from demand centers, posing a challenge in terms of expanded development and deployment. But CSP plants typically include their own heat storage system, which increases the facility's capacity while enabling on-demand electricity and better grid integration of renewable solar power. This could allow the technology to compete with traditional base- or intermediate-load power plants.

Besides providing renewable power, CSP can also generate significant amounts of valuable high-temperature heat for industrial processes, the production of solar-based clean fuels (such as hydrogen and other energy carriers), and water desalination. CSP's unique capabilities of reducing energy losses (through direct conversion of sunlight into thermal energy) and generating heat at a wide range of temperatures (400°C to over 1000°C) make it an efficient and flexible option for industry. In the food and beverage industry, for example, solar thermal heat can be used in nut processing, dairy pasteurization, meat and seafood processing, cooking, and sterilization. For instance, Horizon Nut used Sunvapor's CSP technology to make steam, which was used directly to blanch, pasteurize, and roast pistachios and almonds.[35] The chemicals, plastics, and textile industries also have processes with varying temperature needs, where coupling with solar thermal power can replace fossil fuels.

Wind

The basic concept behind the operation of a wind turbine is the conversion of the kinetic energy of wind into mechanical power via blades, which then drive the wind turbine's generator to create electricity. Wind energy can be harvested on land using large, utility-scale turbines connected to the electric grid, or much smaller turbines that may or may not be connected to the electric grid. Grid-connected turbines are becoming larger and more powerful, with land-based wind turbines typically having capacities of around 3 MW[36] and offshore turbine capacities reaching up to 10 MW.[37] While large wind farms can be located anywhere with good wind resources (though not necessarily near demand centers because the electricity generated can be transmitted via the grid), small wind systems must find locations with both good wind resources and demand centers. As a result, these systems are typically found in rural areas and on farms.

Wind speed and wind quality are the two most important factors to consider when locating a wind turbine (or wind farm). Wind resources are classified into seven classes based on wind speed (m/s) and wind power density (W/m^2). Class 2 areas are marginal for utility-scale wind energy development, while Class 1 areas are unsuitable. The higher the class, beginning with Class 3, the better the wind resource for development. As the amount of available energy in wind is proportional to the cube of the wind speed, each increase in wind speed boosts the amount of energy that can be harvested. For example, doubling the wind speed increases the wind turbine's power output by a factor of eight ($2^3 = 8$). Furthermore, taller turbines can reach higher wind speeds and more consistent winds (because there is less friction from the landscape), resulting in more efficient electricity production.

The roughness of the surrounding terrain affects the wind's force and causes it to slow down. However, in offshore environments, friction is low, allowing for stronger and more consistent winds and thus a greater potential for wind energy harvesting. Moreover, daily offshore wind speed profiles correspond well to periods of high electricity demand by coastal cities. As a result, **offshore wind** energy generation is a rapidly evolving technology. While land-based wind resources have been used for quite some time, the first offshore wind projects were deployed in the 1990s, with significant growth in the years that followed. And as technology develops, offshore turbines are being placed in deeper waters, where the turbine is attached to a floatable structure (tethered to the seafloor) rather than a fixed foundation in the sea, allowing for even greater wind energy capture. However, the logistics and natural conditions at sea are more challenging, necessitating more resilient components, a wider range of sea vessels, and thus higher costs.

There are several examples of novel designs for wind energy harvesting.

The Vortex bladeless wind turbine[38] generates electricity when a slender tower (no blades) vibrates at a specific frequency as the wind blows. The Vortex turbine has a smaller footprint and costs less to produce and maintain than other models. The lack of blades makes it quieter and safer for wildlife, rendering it attractive for both urban and suburban communities. A small, portable version has also been developed, making it easy to transport and set up in remote, off-grid, and disaster-stricken areas.

Typical wind farms require large swaths of suburban land, and transmitting this power to large demand centers such as cities requires investment and reliance on long transmission and distribution lines. Aeromine[39] has developed a design which can be installed on buildings with large flat rooftops, such as warehouses, manufacturing facilities, data centers, big-box retail stores, multi-family residential buildings, and office buildings. The technology can capture wind speeds as low as 5 miles per hour and amplify the flow of air through an airfoil design. The Aeromine system needs only 10% of the roof space typically occupied by solar panels, and when combined with solar panels, can allow for a building to meet 100% of its energy needs while reducing energy storage requirements.

Biome Renewables looked to nature to make wind turbines quieter and more efficient.[40] The motion of typical turbine blades creates air streams with different speeds, and when these streams of air meet behind the blade, they collide, producing both noise and turbulence. Looking at the silent flight of owls, researchers discovered that a specialized fringe along the trailing edge of the owl's wing forces the air to mix at specific locations, thus muffling the sound of flowing air. Biome Renewables used this principle to design wind turbine blades with serrated edges, which reduces noise by 50% to 80% while also increasing power generation because the reduced turbulence allows the blades to spin faster at the same wind speed.

Researchers at Istanbul Technical University in Turkey collaborated with the tech firm Devecitech[41] to develop and deploy vertical wind turbines that could be placed on highways and other transportation corridors to capture the airflow created by passing traffic, along with natural wind, to generate 1 kilowatt of energy per hour. These devices can also be combined with solar PV for extended energy, and sensors to track factors such as temperature, humidity, carbon footprint, and earthquake activity.

Geothermal

The amount of heat energy contained in the Earth's core (geothermal energy) is based on heat from the planet's formation billions of years ago, as well as heat from the radioactive decay of naturally occurring unstable isotopes. In contrast to intermittent renewable resources such as solar and wind, geothermal power plants are considered baseload power plants, which means they can provide a continuous source of power. This is a key feature that distinguishes this renewable resource.

Geothermal heat can be found at a wide range of depths and temperatures, making it a versatile resource. Lower-temperature heat available at relatively shallow depths, for example, can be used to heat and cool homes and commercial buildings using highly efficient geothermal heat pumps. The technology is based on the top 3 meters (10 feet) of the ground being relatively constant in temperature throughout the year, warmer than the air above it in the winter and cooler in the summer, just like a cave. The geothermal heat pump takes advantage of this by transferring heat stored in the earth or in groundwater into a building during the winter and out of the building and back into the ground during the summer. In other words, the ground serves as a heat source in the winter and a heat sink in the summer.

In areas with modest heat flows, the resource could be used as heat for direct industrial applications such as food processing, greenhouses, aquaculture, materials drying, and paper manufacturing, thereby replacing the use of fossil fuels. And high-temperature geothermal resources at greater depths can be used to generate electricity. While low-temperature, shallow resources are widely available, deeper, high-temperature resources are more limited to specific locations. Advanced technologies are being developed globally to extract more of the resource in more diverse locations.

Biomass

There are several advantages to generating electricity from biomass resources. Biopower is one of the few baseload renewable options that can use a wide range of feedstocks. Furthermore, because biomass technologies are highly scalable, they can be used to generate electricity for many applications, from farms and remote villages to large cities.

The economic feasibility of biopower depends on the availability of a reliable, long-term supply of a suitable biomass feedstock at a reasonable cost. Feedstock derived from urban and industrial waste includes wood waste (such as packing crates and pallets), biogas from wastewater treatment, landfill gas from municipal solid waste, and food-processing residues. Rural feedstock sources include forest residues, agricultural residues (such as corn stovers and wheat stalks), fast-growing energy crops (grown specifically for energy production), and biogas produced from livestock effluent. And the main processes used to convert biomass resources into electricity are thermal-chemical processes (combustion, gasification, and pyrolysis) and biochemical processes (anaerobic digestion).

To improve handling, transport, and conversion, biomass feedstock must be pretreated and upgraded. The moisture content of biomass is an important factor to consider, and it can be higher in the case of some organic wastes. When a high moisture content biomass is used, the energy value of the feedstock decreases, raising transportation and fuel costs. If the feedstock is dried, its energy density can be enhanced, which helps reduce transportation costs and improves combustion efficiency. But the tradeoff is that these extra processing steps raise feedstock prices while decreasing the energy balance significantly due to the energy consumption of pretreatment.

Some other concerns with this power production pathway are food security, biological diversity, and water use. To meet the food needs of an expanding population, global food production must increase significantly. The increased cultivation of energy crops will place an increased demand on agricultural products, which will most likely be met by intensifying existing agricultural systems, or claiming new agricultural land, all at the expense of biological diversity. We can reduce these negative impacts while also solving our growing waste problem if we focus more on using residues and waste as feedstock. The water footprint of bioenergy production can also be substantial, and varies based on factors such as climate, crop types, yields, and agricultural practices. Generating 1 gigajoule of electricity from biomass can require about 50–400 m^3 of water.[42]

Integrated biorefineries, as explained in Chapter 2, are an emerging application. These facilities can extract more value from biomass by processing various types of biomass feedstocks into energy and a variety of marketable products such as food, feed, materials, and chemicals. The operation becomes economically competitive by producing high-value byproducts.

Hydropower

For decades, hydropower has been used to generate electricity; it was one of the first applications of renewable energy. When water flows along a river or stream, it has potential energy due to its elevation. As the water descends, this potential energy transforms into kinetic (motion) energy, turning a turbine. The turbine then spins a generator, converting the kinetic energy into electricity.

Hydropower plants can be run-of-river (little or no storage capacity), reservoir (water is stored behind the dam in a reservoir), or pumped storage (off-peak electricity is used to pump water from a reservoir to storage and is then used to generate electricity during peak times). And not all plants need to be large in scale. Micro (5–100 kW) and pico (a few hundred watts up to 5 kW) hydropower can provide power for small communities or remote areas away from the grid.

Because hydropower has inherent storage capabilities and its generation is more predictable, it is not considered variable in the same way that wind or solar power are. However, because it is dependent on precipitation and water runoff, hydropower may be considered variable over longer time scales. Reservoirs, which can regulate water flows for freshwater supply, flood control, irrigation, navigation services, and recreation, also contribute to climate change adaptation efforts.

Marine and Hydrokinetic

Marine energy technologies harvest energy from the natural movement (waves, tides, currents) or temperature/ salinity differences of water bodies and have a very large energy potential, but they are still in the early stages of development and have not yet been widely commercialized. Tidal and wave resources are much more energy dense than solar and wind resources, with waves having the highest energy density. Because of their high energy content profiles, marine energy resources have a lower capital cost per unit of electricity generated, but their operation and maintenance costs are unpredictable and can be high (due to the harsh operating environment at sea), accounting for a significant portion of total project costs.

Tidal projects account for the majority of deployed marine kinetic capacity to date and are farther advanced in terms of technology. Different technologies can be used to harvest energy from either the tidal currents or the difference between high and low tides. Tidal currents are regular, predictable, and unaffected by weather and climate conditions, which is an important advantage. The technology used to harvest tidal energy is like that of conventional wind turbines. Since water is approximately 800 times denser than air, tidal turbines are stronger and heavier than wind turbines and can capture more energy.

Waves form when the sun heats the Earth's surface unevenly, causing winds. The motion in waves contains kinetic energy, which can be harvested to generate clean electricity. Waves vary hourly or seasonally, and estimating their characteristics involves less uncertainty than wind. Harvesting technologies are either fixed to the ocean floor or floated on the ocean's surface, and they can be deployed near the shore or in deeper waters. The energy from wave movement can be captured using a variety of methods (the industry has not converged on a single design yet), such as devices that bend with the waves or devices that use pressure fluctuations caused by wave movement.

Ocean thermal energy conversion (OTEC) technology relies on the temperature difference between the warmer water surface and the cooler water depths. Because a temperature gradient of about 20°C is required, the technology works best in equatorial regions. The warm surface water vaporizes a working fluid with a low boiling point. The vapor then spins a turbine that is coupled with a generator, creating electricity. The vapor is cooled by the colder deep-sea water, condensing back to a liquid and being reused in the system, resulting in a baseload system. Aside from electricity generation, OTEC systems can contribute to a variety of other value-added applications such as desalination, air conditioning, agriculture, aquaculture, and mineral extraction.

The difference in salt concentration, also known as **salinity gradient**, can be used to generate electricity at points where a river flows into the sea. Technologies that use semi-permeable membranes (which only allow specific substances to pass through) are being researched but are still in the early stages of development. This pathway can be used as an additional continuous power source once it is further developed.

Wave and tidal technologies are in direct contact with the marine environment, which poses additional challenges. One source of concern is the accumulation of marine organisms on submerged structures. These problems are mitigated by using some moving parts and specific assemblies. Offshore devices must also be designed to withstand the corrosive nature of the sea. Furthermore, offshore electric distribution networks face more permitting challenges than onshore systems due to regulations relating to fisheries, shipping, oil and gas exploration, conservation areas, and water borders and rights.

Alternative Fuels and Vehicles

Alternative fuels such as compressed natural gas and ethanol have seen an increase in use over the last few decades, but it is barely noticeable in the grand scheme of things. And over the last decade or so, transportation has been shifting to an all-electric platform, with battery, plug-in hybrid, and fuel cell electric vehicles.

Alternative fuels can be made from both fossil and renewable sources. Fossil-based alternative fuels don't require much modification of vehicle technology, and they reduce carbon emissions over their entire lifecycle, but they are subject to price volatility due to fluctuating fossil fuel availability and prices. Advanced vehicle technology and alternative fuel infrastructure must coexist; however, vehicle technology is currently more commercialized than infrastructure. Infrastructure is more complex due to feedstock considerations, as well as logistics for transportation, storage, and fueling stations.

Alternative fuels can be categorized broadly as fossil-, alcohol-, and bio-based, and the more specialized categories of hydrogen and electricity. Fossil-, alcohol-, and bio-based fuels require only minor changes to existing vehicle technology. However, the vehicle must switch to an all-electric or mostly electric platform to be able to use hydrogen or electricity.

Fossil-Based Fuels: Liquefied Petroleum Gas (LPG)

LPG, also known as propane, is a byproduct of petroleum refining or natural gas processing and is commonly used in internal combustion engine (ICE) vehicles. It is a gas at normal temperatures and pressures, but when stored under pressure, transforms into a liquid. LPG is also used in a variety of other applications, including home and water heating, cooking, powering farm and industrial equipment, and drying corn, and as a raw material in the manufacture of plastics and other compounds.

A gallon of LPG contains 73% of the energy of a gallon of gasoline,[43] which means that more fuel is required to travel the same distance, leading to a lower fuel economy. LPG has a lower sulfur content than gasoline, resulting in lower sulfur emissions when used in an ICE vehicle. Other pollutants and GHG emissions are also lower when compared to gasoline-powered vehicles. The infrastructure requirements for an LPG fueling infrastructure are like those of a gasoline infrastructure.

Fossil Based Fuels: Natural Gas

Natural gas can be used in vehicles as compressed natural gas (CNG) or as liquified natural gas (LNG). While CNG is stored in tanks at pressures up to 3600 pounds per square inch, LNG is cooled to -260°F and stored in insulated tanks. Because liquids are denser than gases, more natural gas can be stored in liquid form, allowing vehicles to have longer driving ranges. Natural gas can be used in dedicated natural gas vehicles or dual-fuel vehicles that can run on both natural gas and gasoline or diesel.

The use of natural gas in vehicles reduces smog-forming pollutants by 60% to 90% and GHG emissions by 30% to 40%.[44] LNG stations are similar in structure to gasoline refueling stations because they dispense liquid fuel. However, because the fuel is extremely cold, protective measures and clothing are required when dispensing.

Alcohol-Based Fuels: Methanol

Methanol, also known as wood alcohol, is an alcohol-based alternative fuel that can be produced from a variety of carbon-containing sources, including natural gas, coal, and renewable resources such as wood and landfill gas. Compared to other alternative fuels such as CNG and LPG, it has a higher energy content per volume. Methanol can be used directly, blended (5% to 85%) with gasoline or diesel, or converted to dimethyl ether by dehydrating methanol. Although methanol is a toxic substance, its use as a transportation fuel does not pose major concerns. In the event of contamination, methanol is biodegradable and less toxic to humans than gasoline. But using methanol directly in passenger vehicles would require significant engine and lubrication system changes. Furthermore, cold-start issues may arise, requiring the addition of volatile compounds such as butane and propane to improve engine performance in cold climates.

Alcohol-Based Fuels: Ethanol

Ethanol is a renewable alcohol fuel that can be made from a variety of starch- and sugar-based biomass feedstocks (such as corn and sugarcane) or cellulosic feedstocks (such as dedicated energy crops, wood residues, crop residues, or industrial waste). Sugar-based feedstocks are easier to process and are the preferred feedstock for ethanol production in the United States. Cellulosic feedstocks do not compete for food choices (as starch-based feedstocks do); however, due to their complex chemical nature, it is more difficult to separate the sugars from these resources to convert to ethanol.

Ethanol can be blended with gasoline in percentages ranging from 10% (called E10) to 85% (called E85). While the combustion of ethanol emits carbon dioxide, this is offset by the carbon dioxide captured when plants are grown for use as ethanol feedstock. As a result, net-zero or lower-carbon emissions are attainable. The energy content of ethanol is lower than that of gasoline, resulting in lower fuel efficiency for vehicles that use ethanol. However, the lower-temperature burning of ethanol is gentler on a vehicle's engine, providing less wear-and-tear and thus longer engine life.

Bio-Based Fuels: Renewable Natural Gas (RNG)

RNG can be generated from a wide range of organic waste streams, including wastewater sludge, food waste, and dairy manure. These organic wastes are fed into biodigesters, where microorganisms convert them into biogas and a nutrient-rich slurry that can be used as fertilizer. Further processing of the biogas converts it to RNG, which is chemically identical to conventional natural gas and can be used as an alternative fuel by taking advantage of the existing distribution infrastructure.

Bio-Based Fuels: Renewable Dimethyl Ether (DME)
Renewable DME is a simple molecule that can be made from a variety of renewable feedstocks, including forest and agricultural residues, animal and municipal wastes, wastewater sludge, and the combination of renewable power and carbon dioxide. Because the fuel has a chemical structure like propane, it can be blended with propane and used in existing LPG vehicles or used to replace diesel in engines with minor modifications.

Bio-Based Fuels: Biodiesel
Biodiesel can be made from a variety of resources, including vegetable oils (new or used, such as soybean, rapeseed, cottonseed, or palm oil), animal fats, and recycled restaurant grease, and it can be blended up to 100% with petroleum diesel in most engines after minor modifications. The use of used or recycled oils serves the dual purpose of mitigating waste and generating sustainable fuel, while avoiding the issues associated with using food crops. More importantly, just as ethanol production must shift toward the use of cellulosic materials, biodiesel must shift toward the use of algae for this alternative fuel to develop in a more sustainable manner.

The use of biodiesel reduces air emissions significantly. Furthermore, it has been shown that the reductions in these pollutants are not affected by the type of feedstock used in biodiesel production. While GHG savings are significant, the elimination of sulfur dioxide emissions is also a major benefit, particularly in areas where local air pollution is a concern. However, the use of biodiesel has been shown to either increase or decrease nitrogen oxide emissions depending on the engine type and testing procedure.

Algae-based biodiesel production holds significant promise despite the lack of a current commercial market. Algae, being efficient photosynthetic microorganisms, can generate biomass more effectively than terrestrial plants by using water, carbon dioxide, and sunlight. They grow in non-arable areas using saline or wastewater, thus not competing for agricultural land or food resources. Algae can convert about 6% of solar energy into biomass, outperforming even the most productive crops such as sugarcane. They also produce oil yields up to 20 times higher than soybeans per hectare.[45] Beyond lipids, which are converted into biodiesel, algae also yield proteins and carbohydrates that can be used to create other alternative fuels such as alcohols, diesel, methane, and hydrogen. Moreover, algae could play a crucial role in reducing greenhouse gas emissions by being cultivated near fossil-fuel power plants to absorb emitted GHGs.

Hydrogen

Historically, hydrogen has been used primarily in ammonia production and oil refineries and then expanded into the hydrogenation of edible fats and oils, metallurgy/furnace annealing, and electronics markets. Today, hydrogen is being used as an energy carrier to fuel cars and buses, generate prime and backup electricity, replace batteries in forklifts, power consumer electronics, and serve as a means of energy storage and grid support.

There are several ways to produce hydrogen. Thermal (using feedstock such as natural gas, coal, or biomass), electrolytic (using electricity to split water), and photolytic (using light and microorganisms or specialized semiconductors) processes are the most common. High-temperature heat from nuclear reactors or solar concentrators can also be used to split water at high temperatures.

Hydrogen can be stored as a high-pressure gas, liquid, or in solid materials, and delivered to point of use via trucks, tube trailers, and pipelines. Liquid hydrogen is at very low (cryogenic) temperatures, reaching as low as -253°C, which requires more energy. However, the liquid form is denser, requiring less volume for storage. If the liquid hydrogen is also pressurized (cryo-compressed hydrogen), then even more hydrogen can be stored.

Hydrogen can also be generated onsite at the point of use, eliminating the need for delivery. For example, a hydrogen refueling station can use the existing natural gas pipelines to generate hydrogen from natural gas (via steam methane reforming, a thermal process), or can use grid (or even onsite solar) electricity in an electrolyzer. Hydrogen refueling stations look like gasoline stations and operate similarly to them, having dispensers and taking only a few minutes of refueling to provide the same range as conventional cars.

Once hydrogen is created, stored, and delivered, its most effective use is via a fuel cell. Fuel cells can provide heat and electricity for buildings and electrical power for vehicles and electronic devices. They work like batteries but do not run down or need recharging. They generate electricity as long as there is a constant supply of hydrogen. Fuel cell vehicles use electric motors and require considerably less energy than gasoline vehicles. A fuel cell system can use 60% of the energy in the fuel, whereas a typical gasoline ICE is only 20% efficient.[46]

Electricity

As technology advances and concerns about oil dependence and climate change rise, there is a gradual shift toward electric vehicles. Electric vehicles are divided into five categories: hybrid electric vehicles (HEV), battery electric vehicles (BEV), plug-in hybrid electric vehicles (PHEV), extended-range electric vehicles (EREV), and fuel cell electric vehicles (FCEV). While all these vehicles have an electric platform and make use of batteries, there are a few differences to consider.

By plugging into an electric power source, BEVs and PHEVs use electricity from the grid or a dedicated renewable power source such as solar PV. A BEV, often known as a pure electric vehicle, is powered only by the energy stored in its onboard battery. The basic idea is the same as with any other battery-powered device: the electricity stored in the battery only gets the car so far before it needs to be recharged (by plugging in). An EREV is a BEV with the addition of a small combustion engine, which supplements the battery and extends the range when the battery is fully drained. A PHEV takes advantage of both an electric motor (which uses power stored in a battery and can be recharged by plugging into the grid or by using energy captured from braking) and an ICE (that can use conventional or alternative fuels). An HEV, like a PHEV, has an ICE and an electric motor. The distinction here is that the battery is charged either by regenerative braking or through the ICE (not by plugging in). A FCEV differs from the other alternatives in that it uses hydrogen as a fuel. The chemical energy in hydrogen is converted to electrical energy using a fuel cell, and this electrical energy drives the vehicle's electric motor. When the hydrogen fuel runs out, the FCEV, like normal gasoline cars, must be refueled with hydrogen at hydrogen filling stations.

Consumers are accustomed to and take for granted the driving range of conventional gasoline ICE vehicles and the established refueling infrastructure. As a result, when contemplating alternative transportation technologies, consumers demand at least the same degree of comfort and reliability that they have come to expect from conventional vehicles over the years. Charging of BEVs and PHEVs takes much longer than the few minutes consumers are used to when fueling their ICE vehicles, but faster charging infrastructure and higher-capacity batteries are continuously being developed. Fueling a hydrogen fuel cell vehicle is more comparable to current ICE vehicles, requiring only a few minutes, but developing a refueling infrastructure is more complex. All these technologies will likely be used concurrently, each serving a specific need.

Energy Storage

Electric energy storage is the process of storing energy in kinetic, potential, or chemical forms for later use. Energy storage serves several important functions, including balancing fluctuations in electricity supply and demand, meeting peak electricity needs by using electricity stored during times of lower demand, and avoiding or delaying the need for expansion of the electric grid. To ensure grid reliability, particularly with increased integration of intermittent renewables, additional conventional (and polluting) generating units or non-generation resources such as energy storage must be built or integrated into the existing grid infrastructure.

The amount of energy (kilowatt-hour or megawatt-hour) that can be stored at any given time is referred to as energy storage capacity. Because over-discharging some technologies (particularly lead-acid batteries) reduces their useful lifetime, many storage technologies have limitations on how much stored energy can be used. Charge and discharge rates are power measurements (kilowatt or megawatt) that indicate the rate at which energy is added (charge mode) or removed (discharge mode) from a storage system. In general, for most energy storage technologies, these rates do not remain constant but rather vary with the amount of energy stored and the length of time power has been continuously removed or added to storage. Most energy storage technologies have lower charge rates than discharge rates, and energy storage systems are commonly described in terms of discharge rates.

The performance and costs of energy storage systems are inextricably linked to the service they are to provide. Backup power systems, for example, require immediate (within seconds) response and moderate power output, and can usually handle higher costs (due to the high reliability provided), whereas grid voltage and frequency regulation applications require a response time ranging from seconds to minutes and higher power output. Power must be available for hours for daily load-leveling activities, and the associated costs are determined by the price of peak-load electricity. Investment costs for storage technologies are also highly variable because, except for pumped hydro technologies, these technologies are mostly in precommercial stages with little operating experience.

Pumped Hydro

Pumped hydro systems are an old and mature technology that dates back to the late 1800s. Electricity is stored by using it to power pumps that move water from a lower-level reservoir to a higher-level reservoir. The technology is capable of large-capacity energy storage (hundreds to thousands of megawatts), has a long lifetime (more than 40 years), is approximately 70% efficient, and has shorter response times (only requiring about a minute to go from zero to full output). Stored energy can be released over long periods of time (up to 10 hours). But these systems face challenges due to location-specific constraints, as their siting is dependent on environmental, topographical, and geological conditions, which can be mitigated to some extent by using underground reservoir systems.

Compressed Air Energy Storage (CAES)

CAES systems are also relatively mature. Electricity is used to inject high-pressure air into underground geologic formations, and this high-pressure air is released from underground and used to power natural gas-fired turbines when needed. CAES systems with capacities of a few hundred megawatts can be built and discharged over relatively long periods of time (a few to 24 hours), making them suitable for load leveling (balancing electricity supply and demand). A conventional CAES system has an efficiency of around 50%, but this can be increased by using an advanced gas turbine. CAES units can be operational in about 15 minutes. However, because natural gas is typically used to power the gas turbine, the technique emits carbon dioxide. And as it is most cost effective to locate CAES systems near salt rock deposits, siting options are limited. Above-ground storage in metal vessels or pipes expands the siting options for CAES systems.

Flywheels

A flywheel is a device that stores electrical energy by spinning at high speeds (30,000–40,000 revolutions per minute). As it discharges electricity, its rotation slows, and the spinning mass drives a generator to produce electricity. Flywheels can provide kilowatt- to megawatt-sized power for a limited time (seconds to minutes) with a quick response time (about 15 minutes), making them ideal for use in small grids for backup power, frequency regulation (correcting for short-term changes in electricity use that may affect the stability of the power system), and wind power support.

Batteries

Large-scale rechargeable batteries, such as sodium sulfur (NaS), lithium-ion (Li-ion), and flow batteries can be used in stationary power storage systems for applications such as power quality and load leveling. Furthermore, as PHEVs become more common, their connection to the electric grid will enable the use of their onboard batteries for electricity storage. Batteries have been developed and used primarily in the consumer electronics and power tools industries over the last several decades. However, research has diversified into advanced battery design for use in the transportation sector. This, in turn, has paved the way for the use of batteries in electric grid energy storage applications.

In terms of large-scale grid energy storage, lead-acid and NaS batteries have the most extensive track record. For distributed storage applications such as residential/commercial PV systems, lead-acid batteries, NaS batteries, and large Li-ion batteries are generally the technologies of choice. They can also be used for frequency regulation and in backup power systems. Li-ion batteries offer the highest energy density, cycle efficiency, durability, and the lowest self-discharge (gradual loss of stored charge due to internal chemical reactions, even when the battery is not in use) among these technologies. While smaller Li-ion batteries for portable devices are commercially available at low prices, using this technology for grid energy storage applications is still prohibitively expensive. They cannot be scaled up as easily as other batteries and require enhanced safety and reliability measures.

Flow Batteries

Flow batteries work similarly to traditional solid batteries, but they can store exponentially more energy because the chamber where the electrochemical reaction occurs is connected to large external tanks containing electrically charged liquid. They also function like fuel cells, but the reaction is reversible, and the electrolytes can be reused. Scaling up is relatively simple (by increasing the volume of the electrolyte), and the cost per unit of energy storage is typically lower than for non-flow batteries, making them more appealing for larger applications. Furthermore, flow batteries have high efficiency (over 75%) and long lifetimes (over 10,000 charge-discharge cycles). Another advantage of flow batteries is that they can operate at room temperature, as opposed to the much higher temperatures of other battery types, such as NaS batteries. Flow batteries are also light in weight and have a longer operational life because their main components are made of plastic.

Supercapacitors

Supercapacitors are similar to lithium-ion batteries. In comparison to conventional electrolytic capacitors, the large surface area provided by these systems allows for much higher energy densities. While supercapacitors still have lower energy densities than lead-acid batteries, they can be cycled hundreds of thousands of times and charge and discharge at a faster rate. Supercapacitors have a fast response time and can be used as instantaneous voltage compensators or in tandem with battery storage.

Superconducting Magnetic Energy Storage (SMES)

SMES systems use a coil with many windings of superconducting wire to store and release energy as the current flowing through the wire increases or decreases. To maintain the superconducting properties of the wire materials, the SMES device must be refrigerated, which adds to the device's maintenance costs. Because SMES systems provide short bursts of energy, they can also be used to improve power quality.

Hydrogen

The conversion of low-cost (off-peak) electricity into hydrogen is an emerging pathway for energy storage. Utilities have become more engaged in this process, especially for the storage of off-peak, cheap, and often curtailed wind power. Hydrogen-based energy storage involves using electricity to split water into hydrogen and oxygen via an electrochemical process known as electrolysis. The hydrogen produced can then be used to generate electricity for grid load management using a hydrogen-powered combustion engine or a fuel cell, or it can be used as a transportation fuel for hydrogen fuel cell vehicles. This method may create a business case for renewable power providers if there is also a strong vehicle fuel demand for hydrogen. While the roundtrip efficiency of producing hydrogen via electrolysis and then converting it back to electricity in a fuel cell is lower than that of other storage methods, the amount of energy stored is significant, and a new value-added product (hydrogen as a transportation fuel) can be produced.

Thermal Energy Storage

Concentrating solar power plants have the inherent ability to store thermal energy in a synthetic oil or molten salt, extending power generation for 1–10 hours after sunset. However, unlike other storage technologies that can store electricity from any source, this storage pathway is limited to storing thermal energy produced by the solar field. There are also thermal energy storage devices that can be deployed at the residential and commercial level. For example, modular ice storage systems generate ice during off-peak power periods to power air-conditioning systems during the peak afternoon load times, augmenting or offsetting electric chiller cooling.

Vehicle-Grid Integration

An emerging storage-related approach is at the intersection of advanced transportation and stationary applications. Depending on the growth and success of electric vehicles, there may be an opportunity to use the onboard energy storage of electric vehicles as distributed energy storage available to the electric grid while the vehicles are plugged in and charging. Smart grid technologies that enable time-of-use pricing could incentivize consumers to charge their vehicles when renewable electricity is available at a lower cost.

Microgrids and Smart Grids

Hurricanes and severe weather events have always been a part of life in certain regions and seasons, but because of climate change they are becoming more frequent and intense, and they are now occurring all over the world. As a result, long power outages occur, costing businesses, critical facilities, and communities billions of dollars each year. For example, in 2023, the United States was hit by 28 weather and climate disasters, each resulting in losses exceeding $1 billion.[47] When the traditional, overburdened central electric grid fails, a **microgrid** can meet critical power needs. A microgrid is a self-sufficient and discrete grid that can use distributed energy sources such as solar, wind, and natural gas, as well as provide energy storage. It is distinguished by three key characteristics: self-sufficiency, localization, and intelligence.

While a microgrid is linked to the central electric grid, it can also disconnect from the grid and operate independently, which is known as **islanding**. When the main grid goes down, the microgrid can supply electricity and even heating/cooling (via combined heat and power units) to the demand centers to which it is connected, such as hospitals, military bases, businesses, neighborhoods, college campuses, and so on.

Microgrids are *intelligent* because they use a controller, which is the system's central component. Without the need for human intervention, the controller orchestrates the batteries, each individual distributed energy generator, and connected buildings and facilities.

The microgrid ecosystem can use this orchestration to meet the specific goals of its customers, such as using the cleanest energy, achieving the lowest prices, or ensuring the highest level of electric reliability. Microgrids can also generate additional revenue by providing support services to the main grid, such as frequency control to maintain grid stability and standby power (known as spinning reserves) that is rapidly available to meet demand.

Microgrids have primarily been used on college campuses and by the military to date. Because of this, adoption is not nearly as widespread as it could be, given its immense potential. However, as electric reliability concerns grow and distributed energy prices continue to fall, the rate of installation is increasing and will become more important.

Making the grid *smart* is another way to ensure a reliable and resilient supply of electricity. We need a grid that can automate and effectively manage new needs as nearly everything in our lives becomes digitized and more complex. The key feature of a **smart grid** is its ability for *two-way*, interactive **communication** between the utility and customers. When a grid outage occurs, smart grid technologies can quickly detect and isolate the problem, minimizing damage and preventing much larger blackouts. Emergency electricity rerouting can be done more strategically, allowing for a faster recovery.

Smarter grids are also ushering in a new era of increased and informed customer participation, giving consumers greater control over their energy-related decisions. Old-fashioned mechanical meters are replaced with smart ones in homes and businesses, which act as a conduit between the customer and their energy supplier.

Consumers can see details in real time, such as how much electricity they use (or even how much electricity a specific device uses), when the most energy is used, when the lowest cost of electricity is, and so on, thanks to automated and complex information transfers. And this information is easily accessible through computers or cell phones. This level of awareness and information enables consumers to save money by better managing their energy use and making money-saving decisions such as shifting heavy electricity use to lower-cost electricity generation times. Utility companies also benefit from two-way communication by being able to balance energy loads, avoid peak demand, and prevent blackouts.

Case Studies

Energy Efficiency

Efficiency Measures at Microbrewery:[48] Rising Tide Brewing Co., a microbrewery in Portland, Maine, has boosted energy efficiency through equipment upgrades and process changes. They installed LED lighting, programmed their steam boiler to run continuously, using waste heat for winter heating, and keep their hot liquor tank warm to avoid reheating. They added a high-efficiency air compressor with a 400-gallon storage tank for their canning operations, reducing energy waste by preventing short-cycling. These changes have improved efficiency as production grows.

Geothermal Heat

Direct uses of geothermal heat can support agriculture, industry, and aquaculture through applications such as crop drying, greenhouse heating, milk pasteurization, food dehydration, gold mining, desalination, and fish farming. Geothermal fluid can also be cascaded for efficiency, first powering industrial processes before its lower-temperature heat is used for space heating.

Food Dehydration:[49] An industrial-grade geothermal food dehydrator installed at the Domo de San Pedro geothermal power plant in Nayarit, Mexico, uses residual geothermal energy to dehydrate food. The plant features three independent drying chambers, each capable of processing up to 3000 kg of food per batch, with a total capacity of 9000 kg per day. It processes mango, pineapple, tomato, and jackfruit, but is equipped to handle other food products as well.

Chilling Produce:[50] Coveyou Scenic Farm in Petoskey, Michigan, installed a geothermal heat pump system to efficiently chill produce coolers in summer and provide low-level heating in winter. The system, which uses horizontal piping beneath the farm's parking lot, significantly reduces energy costs – cooling expenses dropped from $1000 to under $100 per month. The farm also benefits from geothermally heated greenhouse floors, allowing year-round cultivation of tender crops.

Geothermal Heat Pumps

Heating and Cooling a Manufacturing Facility:[51] Delta Electronics' Fremont, California facility integrates geothermal heat pumps with solar power and battery storage to achieve net-zero energy use. The system circulates 12,000 gallons of water through 92 miles of underground piping, providing 80% of the building's heating and cooling needs. Designed for efficiency, it operates silently and reliably while showcasing Delta's own energy-saving technologies. The facility consumes 70% less energy than an average LEED-certified building, demonstrating the potential of geothermal solutions in sustainable manufacturing.

Solar and Wind

Energy for an Organic Farm:[52] Cross Island Farms, a 102-acre organic farm in New York, installed a 10-kW wind turbine and a 5.52-kW solar panel system. These systems generate about 16,000 kWh annually, covering the farm's energy needs. Wind power typically peaks in the winter, while solar will produce more energy in the summer, ensuring year-round efficiency and lower costs.

Microgrids

Fuel Cell CHP and Microgrid for Winery:[53] In 2011, Stone Edge Farm, an organic winery in Sonoma Valley, California, implemented a 5-kW fuel cell system to deliver combined heat and power to its estate and vineyard, cutting energy use for irrigation, lighting, and heating its 11,000-gallon lap pool. Stone Edge now operates its own microgrid to provide sustainable energy for facility operations. The microgrid includes 26 kW of fuel cells to help power servers, fiber, Ethernet, gates, alarms, and security lighting. Hydrogen for the fuel cell is generated renewably from solar-power electrolysis and also provides fuel for the winery's multiple fuel cell vehicles.

Microgrid for Eco-Hotel:[54] Hawai'i Island Retreat, a 20-room eco-hotel in North Kohala, operates entirely on renewable energy, including solar, wind, and battery storage. Due to its remote location, grid connection was costly, leading the owners to build an off-grid system instead. Since opening in 2008, they have expanded solar capacity, added a wind turbine, and improved battery storage to meet growing energy needs. The retreat has operated without outages for 15 years, ensuring sustainability and resilience. Between 2015 and 2024, the hotel saved 405,420 kWh of energy, reducing costs by $120,000, and cutting carbon dioxide emissions by 370 metric tons.

Chapter 4

What Tools Can You Use? New Realities

Artificial intelligence, virtual reality, augmented reality, and blockchain are revolutionizing the energy sector, enabling smarter, more immersive, and more transparent solutions that improve how we use and save energy.

Artificial Intelligence (AI)

Machines are increasingly simulating human intelligence and playing a larger role in our lives. AI is transforming energy management by improving resource and demand forecasting, control, and response as renewable integration grows. It enhances grid flexibility and resilience amid climate challenges. Traditional resource forecasting relies on narrow weather models, while AI analyzes historical data, real-time sensors, and satellite images. This level of detail can improve solar forecasting accuracy by up to 30%.[55] AI also refines demand forecasting. Smart meters provide regular consumption data, which AI processes hourly to predict load and energy use habits up to 24 hours ahead. This improves grid flexibility and enables smarter energy storage decisions based on demand, generation, prices, and congestion.

Virtual Reality (VR) and Augmented Reality (AR)

VR and AR can enhance safety and cut costs in utility operations. VR immerses users in a digital environment through headsets and haptic controllers, enabling realistic interaction. AR overlays digital elements onto the real world, adding graphics, videos, and sounds without special equipment or obstructing vision.

VR enhances safety training for grid operations, technology installation, and first responders by providing immersive, risk-free simulations. For example, firefighters can practice various response scenarios in handling fires at alternative fueling stations without real danger. VR allows training to happen offsite, reducing costs compared to transporting employees to physical locations.

AR enhances three-dimensional model creation by converting two-dimensional diagrams into interactive visuals, helping technicians better understand equipment. If an expert isn't onsite, AR-enabled devices can transmit data remotely for guidance. AR gathers information from various sources. After a hurricane, residents can submit damage reports through photos and videos, while unmanned systems provide visual, infrared, and global positioning system data. Before dispatching technicians, these insights help assess damage and plan a response, improving safety and efficiency.

Blockchain

Blockchain is a secure, decentralized online ledger that stores encrypted data across a distributed network. No single party controls it, and data logging prevents corruption. An audit trail verifies transactions without third-party intermediaries, saving time and money. In a blockchain-based smart grid, customers can compare energy providers, buy directly, and adjust choices based on real-time electricity prices. Blockchain also improves transparency for renewable energy certificates – market-based instruments that certify the purchaser owns 1 megawatt-hour of clean electricity – tracking details like time, location, and carbon intensity. This ensures buyers receive verified clean energy and reliable low-carbon attributes.

Chapter 5

Now What? The Path Forward

As a small business leader, you face both challenges and opportunities in today's evolving energy landscape. But the way forward is clear. Sustainability and clean energy are no longer optional – they are key to resilience, innovation, and long-term success. Businesses that act boldly will thrive, cutting costs, improving efficiency, and meeting customer expectations. Your decisions have the power to shape not only your company but also your community and industry. Clean energy isn't just about reducing emissions – it's about building a stronger, more competitive business that drives meaningful change.

The shift toward a net-zero economy is inevitable, and sustainability is becoming ingrained in every aspect of business. Just as digital transformation reshaped industries in the 2010s, sustainability is now driving fundamental change. Companies that fail to adapt risk falling behind. The age of voluntary carbon disclosure is ending – measuring climate risk and emissions is now a tool for preparedness, not just compliance. Larger corporations are prioritizing carbon tracking in their supply chains, giving sustainable small businesses a competitive advantage and increased recognition from key industry leaders.

At the same time, shifting political priorities bring both opportunities and challenges. While tax reforms and deregulation may ease financial pressures, trade and immigration policies create uncertainties that could disrupt operations. Small business leaders must remain agile, informed, and proactive in adapting to these changes. Whether it's rethinking resources, exploring new markets, or investing in innovation, adaptability will be the key to thriving in an evolving economy.

Clean energy choices can seem complex, but the CLEAR framework simplifies the process, equipping you with the tools to make informed decisions that strengthen operations, align with business goals, and drive lasting success. The energy transition isn't just about adding renewables – it's about transforming the entire system to be more efficient, resilient, and adaptable.

Start by shifting your mindset – view energy decisions as opportunities rather than burdens. Prioritize efficiency to reduce costs and prepare your business for rising energy prices and climate risks. Then, integrate clean energy solutions that fit your needs, combining renewables, energy storage, and circular economy principles for flexibility. Most importantly, engage your team and stakeholders to align energy strategies with your vision for sustainability and success.

For small businesses, managing costs is always a priority. The five pillars of the CLEAR framework provide a structured approach to reducing expenses by looking at the bigger picture, beyond individual products. By taking action to lower emissions, you create a leaner, more effective business – one that eliminates waste and strengthens long-term financial stability.

No decision is too small to make an impact. Every efficiency upgrade, clean energy adoption, and mindset shift brings us closer to a sustainable future. Your leadership matters – your decisions shape the economy, communities, and the next generation. Step forward with confidence, lead boldly, and build a business that thrives while driving meaningful change!

Sustainability and clean energy are more than commitments – they are competitive strategies. The CLEAR approach will guide you.

About the Author

Elvin Yüzügüllü is a clean energy and sustainability expert, seamlessly blending technical knowledge with strategic communication to empower stakeholders. With a foundation in environmental engineering, she transitioned into energy management during her graduate studies, gaining a holistic systems perspective. Over the past 20 years, she has guided clients through strategic planning, goal setting, and high-impact decision-making on national energy priorities. She also enjoys teaching graduate-level courses, offering insights into energy sector dynamics and analytical tools for managing facility energy use. Elvin holds a Doctor of Science in Environmental & Energy Management and finds creative inspiration in abstract painting.

Note to Reader

Thank you for reading *Navigating Clean Energy Choices*. I hope this book provided valuable insights and continues to support your clean energy decisions while inspiring new ideas. If you have a moment, I would greatly appreciate your review of the book on your favorite platform. Stay connected by visiting www.navigatingcleanenergy.com, where you'll find updates, deeper insights, and essential resources to support your clean energy journey. Your partner in navigating complexity…

Endnotes

1 World Trade Organization. (n.d.). *Small business and climate change*. https://www.wto.org/english/tratop_e/msmes_e/ersd_research_note3_small_business_and_climate_change.pdf

2 World Economic Forum. (2021). *Net zero challenge: The supply chain opportunity*. https://www3.weforum.org/docs/WEF_Net_Zero_Challenge_The_Supply_Chain_Opportunity_2021.pdf

3 S&P Global Commodity Insights. (2022, March 11). *EU plans proposal to phase out Russian fossil fuels by 2027: Von der Leyen*. https://www.spglobal.com/commodityinsights/en/market-insights/latest-news/electric-power/031122-eu-plans-proposal-to-phase-out-russian-fossil-fuels-by-2027-von-der-leyen

4 The White House. (2022, March 25). *Fact sheet: United States and European Commission announce task force to reduce Europe's dependence on Russian fossil fuels*. https://bidenwhitehouse.archives.gov/briefing-room/statements-releases/2022/03/25/fact-sheet-united-states-and-european-commission-announce-task-force-to-reduce-europes-dependence-on-russian-fossil-fuels/

5 Washington Post. (2022, June 22). *Russia's chokehold over gas could send Europe back to coal*. https://www.washingtonpost.com/world/2022/06/22/coal-plant-europe-germany-austria-netherlands-russia-gas/

6 International Renewable Energy Agency. (2021). *Renewable power generation costs in 2020*. https://www.irena.org/-/media/Files/IRENA/Agency/Publication/2021/Jun/IRENA_Power_Generation_Costs_2020.pdf

7 International Renewable Energy Agency. (2021). *World energy transitions outlook: A pathway to 2050.* https://www.irena.org/-/media/Files/IRENA/Agency/Publication/2021/Jun/IRENA_World_Energy_Transitions_Outlook_2021.pdf

8 U.S. Department of Energy. (2024). *United States energy and employment report 2024.* https://www.energy.gov/sites/default/files/2024-08/2024%20USEER%20FINAL.pdf

9 International Renewable Energy Agency. (2023). *Renewable energy and jobs: Annual review 2023.* https://www.irena.org/Digital-Report/Renewable-energy-and-jobs-Annual-review-2023

10 U.S. Department of Energy. (n.d.). *Assess and reduce plug and process loads in your building.* Better Buildings Solution Center. https://betterbuildingssolutioncenter.energy.gov/assess-and-reduce-plug-and-process-loads-your-building

11 U.S. Energy Information Administration. (2020, July 21). *More than 60% of energy used for electricity generation is lost in conversion.* https://www.eia.gov/todayinenergy/detail.php?id=44436

12 Yale Climate Connections. (2022, October 24). *Energy loss is single-biggest component of today's electricity system.* https://yaleclimateconnections.org/2022/10/energy-loss-is-single-biggest-component-of-todays-electricity-system/

13 U.S. Energy Information Administration. (n.d.). *How much electricity is lost in electricity transmission and distribution in the United States?* https://www.eia.gov/tools/faqs/faq.php?id=105&t=3

14 U.S. Energy Information Administration. (n.d.). *Use of energy explained: Energy use in industry.* https://www.eia.gov/energyexplained/use-of-

energy/industry.php# and U.S. Energy Information Administration. (2016). *International energy outlook 2016*. https://www.eia.gov/outlooks/ieo/pdf/0484(2016).pdf

15 Enerdynamics. (n.d.). *How much primary energy is wasted before consumers see value from electricity?* https://www.enerdynamics.com/Energy-Currents_Blog/How-Much-Primary-Energy-Is-Wasted-Before-Consumers-See-Value-from-Electricity.aspx

16 Compressed Air and Gas Institute. (n.d.). *Working with compressed air: Cost of compressed air*. https://www.cagi.org/working-with-compressed-air?ap4=1_2

17 Ellen MacArthur Foundation. (n.d.). *What is a circular economy?* https://www.ellenmacarthurfoundation.org/topics/circular-economy-introduction/overview

18 Ellen MacArthur Foundation. (n.d.). *To fulfil the Paris Agreement, we need a circular economy*. https://www.ellenmacarthurfoundation.org/articles/to-fulfil-the-paris-agreement-we-need-a-circular-economy

19 Signify. (2019). *Case study: Schiphol Airport Amsterdam*. https://www.assets.signify.com/is/content/PhilipsLighting/Assets/signify/global/20191211-schiphol-case-study.pdf

20 Ellen MacArthur Foundation. (2021, June 18). *Building a world free from waste and pollution*. https://www.ellenmacarthurfoundation.org/articles/building-a-world-free-from-waste-and-pollution

21 Circle Economy. (2019, January 22). *The circularity gap report 2019: Closing the circularity gap in a 9% world*. https://www.circularity-gap.world/2019

22 Ellen MacArthur Foundation. (2021). *Completing the picture: How the circular economy tackles climate*

change. https://www.ellenmacarthurfoundation.org/completing-the-picture

23 Princeton University's Carbon Mitigation Initiative. (n.d.). *Stabilization wedges*. https://cmi.princeton.edu/resources/stabilization-wedges/

24 Salvia, A. L., et al. (2019). An analysis of the applications of Analytic Hierarchy Process (AHP) for selection of energy efficiency practices in public lighting in a sample of Brazilian cities. *Energy Policy, 132*, 854–864. https://doi.org/10.1016/j.enpol.2019.06.021

25 Oliveira, M., Fontes, D., & Pereira, T. (2014). Multicriteria decision making: A case study in the automobile industry. *Annals of Management Science, 3*(1), 109–128. https://digitalscholarship.tnstate.edu/ams/vol3/iss1/6/

26 National Highway Traffic Safety Administration. (2024, June 7). *USDOT finalizes new fuel economy standards for model years 2027–2031*. https://www.nhtsa.gov/press-releases/new-fuel-economy-standards-model-years-2027-2031

27 Energy Information Administration. (n.d.). *Use of energy explained: Energy use in industry*. https://www.eia.gov/energyexplained/use-of-energy/industry.php

28 U.S. Department of Energy. (n.d.). *Combined heat and power basics*. https://www.energy.gov/eere/amo/combined-heat-and-power-basics

29 Canary Media. (2022, June 13). *Why we need to recycle clean energy technologies and how to do it*. https://www.canarymedia.com/articles/clean-energy/why-we-need-to-recycle-clean-energy-technologies-and-how-to-do-it

30 Siemens Gamesa. (n.d.). *Pioneering technology: RecyclableBlade.* https://www.siemensgamesa.com/en-int/explore/journal/recyclable-blade

31 Norton Rose Fulbright. (2022, October 25). *Recycling and renewable energy.* https://www.projectfinance.law/publications/2022/october/recycling-and-renewable-energy/

32 U.S. Department of Energy. (n.d.). *Perovskite solar cells.* https://www.energy.gov/eere/solar/perovskite-solar-cells

33 Ko-Solar. (n.d.). *Solar sound barriers & transportation-based solar projects.* https://www.ko-solar.com/

34 U.S. Energy Information Administration. (2011, March 16). *Concentrating solar power technologies offer utility-scale power production.* https://www.eia.gov/todayinenergy/detail.php?id=530

35 U.S. Department of Energy. (2017, June 16). *EERE success story: Concentrating solar power transforms food processing industry.* https://www.energy.gov/eere/success-stories/articles/eere-success-story-concentrating-solar-power-transforms-food

36 U.S. Department of Energy. (2024, August 21). *Wind turbines: The bigger, the better.* https://www.energy.gov/eere/articles/wind-turbines-bigger-better

37 National Renewable Energy Laboratory. (2020, April). *News stories: Floating wind turbines on the rise.* https://www.nrel.gov/grid/news/program/2020/floating-offshore wind rises

38 Vortex Bladeless. (n.d.). https://vortexbladeless.com/

39 Aeromine Technologies. (n.d.). https://aerominetechnologies.com/

40 Santa Barbara Independent. (2023, July 20). *Biomimicry is improving electrical wind generation.* https://www.independent.com/2023/07/20/biomimicry-is-improving-electrical-wind-generation/

41 EuroNews. (2021, November 4). *Turning traffic into clean energy: How this startup plans to power the streets of Istanbul.* https://www.euronews.com/green/2021/11/04/turning-traffic-into-clean-energy-how-this-startup-plans-to-power-the-streets-of-istanbul

42 Gerbens-Leenes, Winnie, et al. (2009, June 23). The water footprint of bioenergy. *Proceedings of the National Academy of Sciences, 106*(25), 10219–10223. https://www.pnas.org/doi/full/10.1073/pnas.0812619106

43 U.S. Department of Energy. (n.d.). Fuel properties comparison. Alternative Fuels Data Center. https://afdc.energy.gov/fuels/properties

44 Rao, G. L. N., & Ramadhas, A. S. (2011). Compressed natural gas. In A. S. Ramadhas (Ed.), *Alternative fuels for transportation* (pp. 227–242). CRC Press.

45 Darzins, A., Pienkos, P., & Edye, L. (2010, August 6). *Current status and potential for algal biofuels production*. IEA Bioenergy Task 39, Report T39-T2. https://task39.sites.olt.ubc.ca/files/2013/05/IEA-Task-39-Current-Status-and-Potential-of-Algal-biofuels0.pdf

46 U.S. Department of Energy. (2015, November). *Fuel cells fact sheet*. https://www.energy.gov/sites/prod/files/2015/11/f27/fcto_fuel_cells_fact_sheet.pdf

47 National Oceanic and Atmospheric Administration. (n.d.). *Hurricane costs*. https://coast.noaa.gov/states/fast-facts/hurricane-costs.html

48 Brewers Association. (2017). *2017 Sustainability Benchmarking Report*. Brewers Association. https://www.brewersassociation.org/association-news/2017-sustainability-benchmarking-report-released/

49 Richter, A. (2020, September 25). *First industrial-grade geothermal food dehydrator of Latin America installed in Nayarit, Mexico*. ThinkGeoEnergy. https://www.thinkgeoenergy.com/first-industrial-grade-geothermal-food-dehydrator-of-latin-america-installed-in-nayarit-mexico/

50 U.S. Department of Energy. (n.d.). *Geothermal heat pump case study: Coveyou Scenic Farm*. Office of Energy Efficiency & Renewable Energy. https://www.energy.gov/eere/geothermal/geothermal-heat-pump-case-study-coveyou-scenic-farm

51 U.S. Department of Energy. (n.d.). *Geothermal heat pump case study: Delta Electronics*. Office of Energy Efficiency & Renewable Energy. https://www.energy.gov/eere/geothermal/geothermal-heat-pump-case-study-delta-electronics

52 New York State Energy Research and Development Authority. (2011). *Cross Island Farms case study: On-site power – Small wind, solar*. NYSERDA. https://nyserda.ny.gov

53 Benjamin, T. G., Gangi, J., and Curtin, S. (2017). *The business case for fuel cells: Delivering sustainable value* (7th ed., ANL/ESD-17/17). Argonne National Laboratory. https://www.energy.gov/sites/prod/files/2017/09/f36/fcto_business_case_fuel_cells_7th_edition.pdf and Stone Edge Farm Microgrid. (n.d.). *Stone Edge Farm Microgrid*. https://sefmicrogrid.com/

54 U.S. Department of Energy. (n.d.). *Case study: Hawaiʻi Island Retreat*. Better Buildings Solution

Center. https://betterbuildingssolutioncenter.energy.gov/sites/default/files/attachments/Hawai%CA%BBi%20Island%20Retreat%20Case%20Study.pdf

55 Mortier, T. (2020, November 24). *Why artificial intelligence is a game-changer for renewable energy.* ReGlobal. https://reglobal.org/why-artificial-intelligence-is-a-game-changer-for-renewable-energy/

Endnotes

CHANGEMAKERS
BOOKS

Transform your life, transform our world. Changemakers Books publishes books for people who seek to become positive, powerful agents of change. These books inform, inspire, and provide practical wisdom and skills to empower us to write the next chapter of humanity's future.

www.changemakers-books.com

Current Bestsellers from Changemakers Books

Resetting Our Future: Am I Too Old to Save the Planet? A Boomer's Guide to Climate Action
Lawrence MacDonald

Why American boomers are uniquely responsible for the climate crisis – and what to do about it.

Resetting Our Future: Feeding Each Other Shaping Change in Food Systems through Relationship
Michelle Auerbach and Nicole Civita

Our collective survival depends on making food systems more relational; this guidebook for shaping change in food systems offers a way to find both security and pleasure in a more connected, well-nourished life.

Resetting Our Future: Zero Waste Living, The 80/20 Way The Busy Person's Guide to a Lighter Footprint
Stephanie J. Miller

Empowering the busy individual to do the easy things that have a real impact on the climate and waste crises.

The Way of the Rabbit
Mark Hawthorne

An immersion in the world of rabbits: their habitats, evolution and biology; their role in legend, literature, and popular culture; and their significance as household companions.